INDIEN
INDIA

Faszinierende Tiere, Kultur und Menschen
Fascinating Animals, Culture and People

Fotos von:/Pictures by:
Jens Beck
Katrin Beck
Frank Hanel
Andreas Klotz
Ingo Knoll
Raghunandan Kulkarni
Harry P. Lux
Harald Lydorf
Michael Matschuck
Martin Ruhnke
Peter Scheufler
Susanne Scheufler
Kerstin von Splényi
Lothar Spranger
Heike Zachenhuber
Norbert Zachenhuber

Vielen Dank für die Unterstützung dieses Projektes an:
Many thanks to the organisations below for supporting this project:

The art of adding value

Indien ist ein riesiges Land voller Gegensätze mit vielen unterschiedlichen Regionen und Menschen. In den auf dieser Karte eingezeichneten Bundesstaaten, vor allem in den rot markierten Orten und fünf grün markierten, großartigen Nationalparks sind sämtliche Bilder für dieses Buch entstanden.

India is a huge country full of contrasts, with diverse regions and people. The photographs in this book were all taken in the states shown on this map, predominantly in the locations marked red and in the five incredible national parks shown in green.

IMPRESSUM/PUBLICATION DETAILS

Indien · Faszinierende Tiere, Kultur und Menschen
India · Fascinating Animals, Culture and People
1. Auflage/First edition, November 2014
ISBN 978-3-9439691-0-8

Idee und Konzeption/Idea and design: Andreas Klotz, Michael Matschuck
Texte/Text: Harald Lydorf, Harry P. Lux, Kerstin von Splényi
Übersetzungen/Translation: Claire Annis, Stuart Potter
Herausgeber/Editor: Andreas Klotz
Layout und Karte/Layout and map: Michael Hildebrand
Fotos/Pictures: Jens Beck, Katrin Beck, Frank Hanel, Andreas Klotz, Ingo Knoll, Raghunandan Kulkarni, Harry P. Lux, Harald Lydorf, Michael Matschuck, Martin Ruhnke, Peter Scheufler, Susanne Scheufler, Kerstin von Splényi, Lothar Spranger, Heike Zachenhuber, Norbert Zachenhuber.
Druck/Printed by: druckpartner Druck- und Medienhaus GmbH, Essen; www.druck-partner.de
Materialien/Materials:
Inhalt/Content: MagnoSatin 170 g/m², Sappi Deutschland GmbH
Schuber/Slipcase: Algro Design 450 g/m², hochwertiger Zellstoff von Sappi/superior paperboard by Sappi
Bezugsstoff/Cover: Velvet Edition Meru, Konrad Hornschuch AG
Verlag/Published by: Fotografeneditionen.de, c/o TiPP 4 GmbH, Von-Wrangell-Str. 2, 53359 Rheinbach, Tel. 02226 911799, www.fotografeneditionen.de, www.tipp4.de

© Copyright 2014 by TiPP 4 GmbH, Rheinbach
Alle Rechte vorbehalten/All rights reserved

Der ganze oder teilweise Abdruck und die elektronische oder mechanische Vervielfältigung gleich welcher Art sind nicht erlaubt. Abdruckgenehmigungen für Fotos und Texte in Verbindung mit der Buchausgabe erteilt ausschließlich die TiPP 4 GmbH. No part of this publication may be reproduced or distributed electronically or mechanically in any form. Permission to reproduce pictures and text in connection with this edition are obtainable exclusively from TiPP 4 GmbH.

Gedruckt mit bluegreenprint® – ein Markenzeichen für nachhaltig ökologisches und soziales Drucken. bluegreenprint® beinhaltet alle geltenden Standards der Druck- und Medienbranche, ist klimaneutraler Druck und fördert weltweit wichtige ökologisch-soziale Projekte. Printed using bluegreenprint® – a hallmark of sustainable ecological and social printing. bluegreenprint® is climate-neutral printing that complies with all printing and media industry standards and fosters important socio-ecological projects worldwide.

*„Ich behaupte, dass ein Geschöpf,
je hilfloser es ist, um so mehr Anspruch hat,
vom Menschen vor der Grausamkeit des Menschen
beschützt zu werden."*

Mohandas Karamchand, genannt Mahatma Gandhi (1869-1948),
indischer Rechtsanwalt, Führer der indischen Befreiungsbewegung

*„I hold that the more helpless a creature,
the more entitled it is to protection
by man from the cruelty of man."*

Mohandas Karamchand, known as Mahatma Gandhi (1869-1948),
Indian lawyer, leader of the Indian Independence movement

Inhalt | Contents

22 Land & Leute / Country & People

Streiflichter durch Alltagsleben, Kultur und Brauchtum

A brief insight into everyday life, culture and customs

48 Ranthambore National Park

Der Tiger ist wieder Herr in seinem Land

The tiger returns to dominance

70 Kaziranga National Park

Weltnaturerbe mit über 100 Jahren Tradition

World Heritage Site with a history that goes back more than 100 years

88 Gir National Park

Heimat der letzten Asiatischen Löwen

Home of the last Asiatic lions

108 Kanha National Park

Wogendes Grasland und „europäische" Wälder

Undulating grasslands and „European" forests

Inhalt | Contents

Bandhavgarh National Park
Im Reich des Bengalischen Tigers
Home of the Bengal tiger

126

Bengalische Tiger
The Bengal tiger

146

Asiatische Löwen
The Asiatic lion

154

Panzernashörner
The Indian rhinoceros

160

Überblick/Karte | Overview/map	3
Impressum | Publication details	6
Grußworte | Forewords	12
Editorial | Editorial	18
Hilfsprojekt | Aid project	166
Fotografen-Viten | Photographer profiles	170
Bildnachweise | Photo credits	174

Grußwort | Foreword

Nach mehreren großartigen Büchern und Kalendern des Mondberge-Projektes liegt nun ein weiterer, atemberaubender Fotobildband vor: „INDIEN – Faszinierende Tiere, Kultur und Menschen". Sollte Ihnen das Mondberge-Projekt noch nicht bekannt sein, dann haben Sie sich mit dem Kauf dieses Buches dazu entschieden, Teil einer kleinen, aber überaus effektiven Umweltschutz-Initiative zu werden: Der Erlös aller Produkte fließt in Hilfsprojekte zum Erhalt aussterbender Tierarten.

Als leidenschaftlichem Dokumentar-Filmer und -Zuschauer erscheinen mir Bücher wie das vorliegende unersetzlich, um immer mehr Menschen die Augen zu öffnen und ihnen zu zeigen, was uns und unserem Planeten verloren geht, wenn wir dem rapide zunehmenden Artensterben weiterhin tatenlos zusehen. Wir machen es der Natur und ihren Arten immer schwerer, den Kampf ums Überleben zu gewinnen. Immer weiter und schneller dringen wir in die natürlichen Lebensräume von Pflanzen und Tieren ein. Artenschutzprojekte wie das Mondberge-Projekt, helfen den Einklang zwischen Mensch und Umwelt zu verbessern oder wiederherzustellen.

Ich wünsche Ihnen einen großartigen Streifzug durch den Subkontinent Indien. Die fantastischen Fotografien von Kultur, Menschen, Natur und den leider selten gewordenen Tierarten dieses Landes wie Tigern, Asiatischen Löwen oder Panzernashörnern werden Sie beim Blättern auf eine unvergessliche Abenteuerreise mitnehmen.

Ihr Hannes Jaenicke

This breathtaking coffee-table book „INDIA – Fascinating Animals, Culture and People" is the latest addition to the outstanding series of books and calendars created by the Mondberge-project. If you are unfamiliar with the Mondberge-project, I'm sure you will be happy to learn that by purchasing this book you have become part of a small but very effective environmental protection initiative. The proceeds from all products sold go towards aid projects for the protection of endangered animals.

As an avid maker and viewer of documentaries myself, I consider books like this to be vital to raising awareness and showing people what will be lost to our planet if we continue to stand by and watch as more and more species become extinct at an accelerating rate. We are making it increasingly difficult for nature, in all of its diversity, to win the fight for survival. We are encroaching ever further and faster on the natural habitat of plants and animals. Wildlife conservation initiatives such as the Mondberge-project help to improve or restore the balance between humans and their environment.

I hope you enjoy this fantastic journey across the Indian sub-continent. The magnificent pictures of the country's culture, its people, nature and, unfortunately, ever rarer animals, such as Bengal tigers, Asiatic lions, and Indian rhinos, will take you on an unforgettable adventure.

Best regards, Hannes Jaenicke

Grußwort | Foreword

Meine Begeisterung über diesen neuen Fotobildband ist riesig. Nach der „Perle Afrikas" gelingt dem Fotografen-Team wieder die Faszination des Publikums. Sie fesseln mit einzigartigen Tierfotos und bezaubernden Aufnahmen der Bewohner des indischen Subkontinents. Eine gelungene Mischung von Natur und Kultur in einer anderen Welt macht die hervorragende Ausstrahlung dieser Fotografenedition aus.

Indien ist mit 1,2 Milliarden Menschen das zweitbevölkerungsreichste Land der Erde und der siebtgrößte Staat. Seine Pflanzenwelt reicht von Hochgebirgsvegetation im Himalaya bis zu tropischen Regenwäldern im Süden. Dank seiner Landschaftsvielfalt bietet Indien eine äußerst artenreiche Tierwelt, viele Arten kommen allerdings nur noch in Rückzugsgebieten vor. Mit der Einrichtung von Schutzgebieten hofft man, bedrohte Arten wie die Bengalischen Tiger, Indische Elefanten, Asiatische Löwen und Panzernashörner, vor dem Aussterben zu retten. Derartige Naturschönheiten für zukünftige Generationen zu erhalten, geht uns alle an. Indien ist ein Land der Gegensätze mit einzigartigen Naturschätzen, aber auch großen Umweltproblemen, z. B. der Wasserknappheit. In den Metropolen Kalkutta, Delhi und Mumbai ist der hohe Gehalt an Feinstaub bedenklich und Plastikmüll findet man in Flüssen und Dörfern.

Die Deutsche Umwelthilfe und Rapunzel unterstützen mithilfe des Hand-in-Hand-Fonds und Partnern vor Ort Projekte, wie z. B. „Sauberes Trinkwasser dank Solarstrom", „Gentechnikfreies Saatgut – Ökolandbau", „Überschwemmungsschutz durch Wiederansiedelung von Mangroven". Wir können dazu beitragen, dass uns die Faszination Indien erhalten bleibt.

Erika Blank
Koordinatorin Hand-in-Hand-Fonds der Deutschen Umwelthilfe
Coordinator of Deutsche Umwelthilfe Hand in Hand Fund

I am incredibly excited about this new coffee-table book. In this latest instalment, following „Perle Afrikas" (The pearl of Africa), the team of photographers have once again tapped the public's imagination. Their unique animal photos and captivating pictures of the people of the Indian sub-continent are spellbinding. A perfect synthesis of the nature and culture of another world makes this coffee-table book a simply outstanding edition.

With 1.2 billion people, India is the second most populous nation on earth and the seventh largest country. Its flora ranges from high-altitude vegetation in the Himalayas to tropical rainforests in the south. And, thanks to its diverse landscapes, India's animal kingdom is extremely varied, although many species now only live in remote regions. The hope is, that by creating protected areas, endangered species like the Bengal tiger, the Indian elephant, the Asiatic lion, and the Indian rhino can be saved from extinction. It is in all of our interests to preserve these wonders of nature for future generations. India is a land of extremes, where unique natural beauty stands side by side with major environmental problems, such as the scarcity of water. The metropolises Calcutta, Delhi and Mumbai suffer from high levels of pollution and the rivers and outlying villages are plagued by plastic waste.

Deutsche Umwelthilfe (German Environmental Relief Agency) and natural food producer Rapunzel support projects such as „Clean drinking water powered by solar energy", „GM-free crops – eco-farming", and „Flood prevention with the resettlement of mangroves" with the aid of the Hand in Hand Fund and local partners. For our part, we can help to keep the fascination of India alive.

Editorial | Editorial

Team 1:
Frank Hanel, Jens Beck, Ingo Knoll, Pushkar P. Achyute, Kerstin Beck, Harald Lydorf, Michael Matschuck.

Im März und im Mai 2014 sind wir in zwei Teams nach Indien geflogen. Wieder einmal sollte es mehr als „nur" eine Fotoreise werden. Unser Ziel ist es, journalistisch zu dokumentieren, eine breite Öffentlichkeit zu informieren, zu unterhalten, aufzuklären und zu begeistern – um damit Bewusstsein zu schaffen – und so jetzt – nach Berggorillas in Uganda, Geparden in Namibia und Großen Soldaten-Aras in Costa Rica – nun auch die letzten Tiger schützen zu helfen! Nicht mehr – und nicht weniger! Weitere Informationen dazu und vieles mehr finden Sie auf **www.mondberge.com**.

Die Reisen durch Indien fanden ohne Probleme statt. Die Organisation durch Eaglerayreisen und die Betreuung von Pushkar und Raghu vor Ort war hervorragend. Wir konnten insgesamt über 50.000 Bilder „schießen". Davon sind nach der Auswahl durch die Fotografinnen und Fotografen etwa 5.000 übrig geblieben und schließlich knapp 400 bei TiPP 4 für diesen Bildband ausgewählt worden. Das erste Treffen beider Teams fand Anfang August 2014 statt. Wir haben Layoutentscheidungen gefällt und die endgültige Bildauswahl getroffen. Es sind genau 272 Bilder in diesem Bildband abgedruckt worden.

Wir würden uns sehr freuen, wenn Ihnen das vorliegende Buch gefällt. Vielleicht möchten Sie ja noch eins verschenken oder es jemand anderem weiterempfehlen? Mit jedem verkauften Buch erhöht sich die Summe, die wir spenden.

Auf diesen 176 Seiten finden Sie aber nicht nur viele Bilder. Wir stellen Ihnen mit einleitenden, aufwendig recherchierten Texten in insgesamt neun Kapiteln zuerst **Land & Leute** vor. Anschließend nehmen wir Sie mit in fünf der schönsten Nationalparks Indiens: **Ranthambore, Kaziranga, Gir, Kanha** und **Bandhavgarh.** Und last but not least werden in drei Kurzkapiteln die beeindruckenden und stark bedrohten Tierarten **Tiger, Löwen** und **Panzernashörner** vorgestellt.

Im Rahmen des Mondberge-Projekts gibt es außer diesem Fotobildband noch jährlich den edlen, großformatigen Wandkalender „Artenschutz", und bisher drei weitere Titel in der Fotografeneditionen-Reihe: „Perle Afrikas – Impressionen aus Uganda", „Afrikas Süden" und „Mittel- und Südamerika – die Natur-Highlights".

Editorial | Editorial

Our latest project saw us fly to India in two teams in March and May 2014, once again with the intention of going on more than „just" a photo trip. Our aim is to educate, entertain, explain and enthuse a wide public audience with our journalistic style and thus raise awareness. Now, after the mountain gorillas of Uganda, the cheetahs of Namibia, and the great green macaws of Costa Rica, this latest book is dedicated to protecting the last remaining tigers. Nothing more, and nothing less! For more information on this and many more subjects, visit **www.mondberge.com**.

We encountered no problems during our travels across India. Our trips were excellently organised by the Eagleray travel company and supported by Pushkar and Raghu on site. We took more than 50,000 photos in total, which the photographers longlisted to around 5,000. These were then whittled down by TiPP 4 to a shortlist of just under 400. The teams met up for the first time in early August 2014 to decide on the layout and finalise the picture selection. Exactly 272 photos were selected to appear in this edition.

We very much hope you enjoy browsing through it. Maybe you would like to give the book to someone as a gift or recommend it. The amount we donate increases with every book purchased.

Not only will you find a wide selection of pictures over these 176 pages, but also thoroughly researched introductory texts divided into nine chapters, starting with „**The country and the people".** In the subsequent five chapters we take you to India's five most beautiful national parks, **Ranthambore, Kaziranga, Gir, Kanha** and **Bandhavgarh**. Last, but not least, three short chapters give you an insight into the impressive yet highly threatened animals, the **Bengal tiger,** the **Asiatic lion** and the **Indian rhino.**

The Mondberge-project also produces a large format wildlife conservation wall calendar each year called „Artenschutz", and has published three more coffee-table books in the Fotografeneditionen series entitled „Perle Afrikas – Impressionen aus Uganda" (The pearl of Africa – impressions of Uganda), „Afrikas Süden" (The south of Africa), and „Mittel- und Südamerika – die Natur-Highlights" (Central and South America – highlights of nature), currently only available in German.

Team 2:
Peter Scheufler, Andreas Klotz, Lothar Spranger, Harry P. Lux, Kerstin von Splényi, Raghunandan Kulkarni, Susanne Scheufler, Heike Zachenhuber, Norbert Zachenhuber.

Land & Leute

Streiflichter durch Alltagsleben, Kultur und Brauchtum

Country & People

A brief insight into everyday life, culture and customs

Land & Leute | The country and the people

Auch wenn die Besuche der fünf Nationalparks und insbesondere die Tierfotografie Hauptgrund unserer Reisen waren, so kam der Kontakt mit den Menschen nicht zu kurz und zumindest ein kleiner Blick auf die vielfältige, oftmals religiös geprägte Architektur war beeindruckend.

Betrachtet man die gesamte Größe des (geschichtlichen) Kulturgebietes, dann mag die von uns besuchte Region zwischen Agra und Khajuraho wie ein Wassertropfen in der Badewanne wirken – aber auf diesem kleinen Gebiet befinden sich vier Weltkulturerbe und neben vielen weiteren Monumenten auch die Paläste von Orcha, Nahtstelle zwischen Mogulreich und Hindu-Königtümern.

Der Mogulkaiser Akbar (1556 bis 1605), der dritte seiner Herrscherlinie, hat neben vielen anderen Bauwerken in seiner Regierungszeit den Bau von drei Weltkulturerbestätten initiiert. Dazu kommen dann noch das von seinem Enkel Shah Jahan erbaute „Rote Fort" in Delhi und das „achte" Weltwunder, der Taj Mahal – eine beeindruckende Epoche!

Khajuraho

Zwischen ca. 950 und ca. 1120 erbauten die Herrscher der Chandella-Dynastie, ein Rajputen-Clan, im kleinen Ort Khajuraho etwa 80 Tempel. Der Ort versank bald in der Vergessenheit, und so wurden die Tempel von den Kriegszügen der späteren Mogulkaiser verschont. Sie wurden 1840 von den Briten in dem damals rund 300 Einwohner zählenden Dorf „wiederentdeckt". Im 20. Jahrhundert durchgeführte Erhaltungs- und Renovierungsarbeiten gestatten heute den Besuch von etwa 20 Tempeln.

Götterfiguren und erotische Szenen in einem Fries des Lakshmana-Tempel, Khajuraho.

Deities and erotic depictions decorate Lakshmana Temple, Khajuraho.

Die Tempel sind im Außenbereich ungewöhnlich reich mit teils sehr erotischen Darstellungen geschmückt. Götterfiguren und „schöne Mädchen" in Posen wie „Wimpern tuschen", „Füße bemalen", „Spiegel benutzen" – immer vollbusig mit eng anliegender dünner Kleidung zeigen ein anderes Verhältnis zu Sinnesfreuden als heute. Man geht davon aus, dass diese Figuren und auch die Positionen der Liebespaare ihren Ursprung im Tantrismus haben. Vereinfacht ausgedrückt: Die Realität ist das Ergebnis der Symbiose

Ort	Monument	Erbauer	entstanden	Klassifizierung
Khajuraho	Tempelbezirk	Chandella-Dynastie	950 -1120	Weltkulturerbe
Agra	Rotes Fort	Mogulkaiser Akbar	1565	Weltkulturerbe
Fatehpur Sikri	Palastanlage	Mogulkaiser Akbar	1571	Weltkulturerbe
Orcha	Palast/Fort	Bundella-Dynastie	1501	---
Agra	Taj Mahal	Mogulkaiser Shah Jahan	1631	Weltkulturerbe

Land & Leute | The country and the people

While the main objective of our trip was to visit the five national parks and to photograph the wildlife in particular, we also had a lot of contact with the people and were impressed by our – albeit – brief look at the country's wonderfully diverse architecture, often influenced by religious beliefs.

Compared with the full scale of the historical and cultural area, the region we visited between Agra and Khajuraho may seem like just a grain of sand in the desert, but this small area boasts four World Heritage Sites and among many other monuments also the palaces of Orchha, where the territory of the Mughal Empire meets the Hindu kingdoms.

Three of the World Heritage Sites in this area were commissioned by Mughal Emperor Akbar (1556-1605), the third ruler of the Mughal Dynasty, along with many other structures built during his reign. This impressive era also produced the Red Fort in Delhi, commissioned by his grandson Shah Jahan, and the „Eighth Wonder of the World", the Taj Mahal.

Khajuraho

The rulers of the Chandella Dynasty, a Rajput clan, built 80 or so temples in the small town of Khajuraho between around 950 and 1120. The town was soon forgotten, which meant the temples were left unscathed by the military expeditions of the Mughal emperors in later times. They were „rediscovered" in 1840 by the British in the village, which was home to around 300 inhabitants at the time. Maintenance and repair work carried out in the 20th century has made it possible to open up some 20 of the temples to tourists.

The outer areas of the temples are exceptionally richly adorned with – in some cases very erotic – depictions. Deities and „beautiful girls" in poses such as „applying mascara", „painting feet", „looking in the mirror" – always full-breasted with closely fitting, thin clothing – are evidence of a different attitude towards sensual pleasures than is the norm today. It is generally assumed that these figures and the positions the lovers are in are Tantric in origin. Put simply: Reality is the result of the symbiosis of the male (= Shiva) and female (= Shakti) principles. The Lakshmana Temple with the Parvati Temple and the figure „man stabbing lion", the mighty double temple of Kandariya Mahadeo and Devi Jagadambi, and the Parsvanath Jain temple are just a few examples.

Agra – Agra Fort

In 1565 Mughal Emperor Akbar had a fort built on the south bank of the Yamuna river, the brick walls of which were clad in red sandstone. The surrounding walls, approximately 1½ miles long and 70 ft high in places, form the outline of a half

Ein neugieriges Streifenhörnchen auf dem weitläufigen Gelände des „Roten Fort" in Agra.

An inquisitive chipmunk on the spacious grounds of Agra Fort.

Land & Leute | The country and the people

von männlichem (= Shiva) und weiblichem (= Shakti) Prinzip ... stellvertretend für alle anderen genannt seien der Lakshmana Tempel mit Parvati Tempel und der Figur „Mann erdolcht Löwe", der mächtige Doppeltempel Kandariya Mahadeo und Devi Jagdambe sowie der Jain Tempel Parsvanat.

Agra – Rotes Fort

Auf einem Hügel am südlichen Ufer der Yamuna ließ der Mogulkaiser Akbar 1565 eine Festungsanlage errichten, deren Ziegelsteinmauern mit rotem Sandstein verkleidet wurden. Die Ummauerung, bis zu 21 Meter hoch, umschließt mit etwa 2,4 km Länge einen halbmondförmigen Grundriss. Der Bau war bereits 1571 abgeschlossen und nur zwei Öffnungen, das Delhi- und Lahore-Tor, gewährten Einlass. Akbar ließ nur mit rotem Sandstein bauen, verlegte aber bereits 1572 seine Hauptstadt nach Fatehpur Sikri. Unter Shah Jahan wurden weitere prächtige Paläste und die berühmte Perlmoschee innerhalb der Mauer gebaut, nunmehr aber mit weißem Marmor verkleidet. Eine Bauweise, die beim Kernbau des Taj Mahal auch angewandt wurde. 1658 wurde das Rote Fort zum Gefängnis für Shah Jahan. Sein Sohn ließ ihn einsperren mit Blick auf das nur 2,5 km entfernte Grabmal seiner Gattin Mumtaz – das Taj Mahal.

Fatehpur Sikri

Akbar hatte lange Zeit keinen Thronfolger – obwohl er eine fleißige Heiratspolitik auch mit Hinduprinzessinnen betrieb – und besuchte deshalb 1568 den 35 km von Agra entfernt lebenden Sheikh Salim Chisti, einen Sufi-Heiligen, der ihm die Geburt eines Knaben vorhersagte. Akbar startete aus Dankbarkeit 1571 den Bau seiner neuen Hauptstadt auf dem Bergrücken als einheitlich geplantes Areal, umgeben von einer gewaltigen Festungsmauer.

Die Moschee Jama Masjid und das Mausoleum von Salim Chisti sind intensiv besuchte Kultstätten, wesentlich ruhiger geht es in der bereits 1586 wieder verlassenen, aber gut erhaltenen, Palastanlage zu. Sehr beeindruckend ist der Weg über den Pachisi Hof (hier wurde mit lebenden Figuren Schach gespielt!) entlang des Wasserbeckens (Anup Talao) zum Diwan-I-Khas. Heute wird gestritten, welche Aufgabe dieses Gebäude hatte, es wird vielfach als Audienzhalle des Kaisers erklärt. Auf dem Weg passiert man den Panch Mahal, ein offener fünfstöckiger Palast, der auf Säulen stehend sich nach oben verjüngt. Die erste Decke wird noch von 84 Säulen getragen, der oberste Pavillon ruht nur noch auf vier. Akbar verlegte seine Hauptstadt etwa 15 Jahre später aus strategischen Gründen nach Lahore (andere Quellen bezeichnen Wassermangel als Grund für die Aufgabe).

Eins der vielen Tore im Eingangsbereich des „Roten Fort" in Agra.

One of the many gates at the entrance to Agra Fort.

moon. Construction was completed in 1571 and there were only two entrances, the Lahori Gate and the Delhi Gate. Akbar only ever built with red sandstone, but by 1572 he had already decided to relocate his capital city to Fatehpur Sikri. Further magnificent palaces and the famous Pearl Mosque were built within the walls of Agra Fort under the reign of Shah Jahan, this time clad in white marble. The same construction method was used for building the core of the Taj Mahal. In 1658, Agra Fort became Shah Jahan's prison. His son had him locked away, facing the Taj Mahal, the mausoleum he had built for his wife Mumtaz, just 1½ miles away.

Fatehpur Sikri

For many years, Emperor Akbar had no heir – although he had plenty of matrimonial alliances, including with Hindu princesses – and so in 1568 he went to visit Sheikh Salim Chishti, a Sufi saint living some 20 miles from Agra, who predicted that a son would be born to him. In 1571, out of gratitude, Akbar started construction of a new capital city on the nearby ridge as a uniformly planned city surrounded by huge, fortified walls.

The Jama Masjid mosque and Salim Chishti's tomb are places of worship that attract a great many visitors, while the palace and its grounds, which were already deserted in 1586 but are well preserved, are far more peaceful. Most impres-

Erdgeschoss des „Panch Mahal", Fatehpur Sikri.

The ground floor of the Panch Mahal, Fatehpur Sikri.

sive is the walk across the Pachisi court (where they used to play chess, using people as the pieces!), alongside the ornamental pool (Anup Talao) to the Diwan-I-Khas. There is some dispute today as to the purpose of this building; many consider it to have been the emperor's private audience hall. On the way, visitors pass the Panch Mahal, a palace of five open storeys of decreasing size. The first ceiling is supported by 84 columns and the topmost pavilion sits upon only four. Around 15 years later, Akbar again relocated his capital city, this time to Lahore, for strategic reasons (other sources cite a lack of water as the reason for giving the city up).

Orchha

The capital city of the Bundella Dynasty, Orchha, also boasts a connection to the Mughal emperors. Not only did Akbar's

Place	Monument	Built by	Built	Classification
Khajuraho	Temple area	Chandella Dynasty	950 -1120	World Heritage Site
Agra	Agra Fort	Mughal Emperor Akbar	1565	World Heritage Site
Fatehpur Sikri	Palace complex	Mughal Emperor Akbar	1571	World Heritage Site
Orchha	Palace/Fort	Bundella Dynasty	1501	---
Agra	Taj Mahal	Mughal Emperor Shah Jahan	1631	World Heritage Site

Orcha

Auch die Hauptstadt der Bundella-Dynastie, Orcha, hat eine Verbindung zu den Mogulkaisern. Nicht nur, dass Akbars Sohn Salim bereits 1594 als Kronprinz den Feldzug zur Eroberung von Orcha befehligte, er freundete sich auch mit dem Herrscher Vir Singh Deo an. Dieser baute das Fort von Orcha weiter aus und zur Erinnerung wurde der neue Palast Jahangir Mahal genannt – Jahangir war Salims Name als Kaiser. Jahangir Mahal ist symmetrisch aufgebaut und beeindruckt durch seine vielen zartgliedrigen Fenster – dem Einfluss der Mogularchitektur zu verdanken.

Das „Fort" liegt auf einer Insel, die durch eine Brücke mit dem Festland verbunden ist. Vielfach wird auch von einer Palastanlage gesprochen, denn die Gebäude, die die Anlage bilden, hatten nicht nur wehrhafte Aspekte. Der Palast der Könige – Raj Mahal – besticht durch seine vielen Deckengemälde. Der dritte Palast in diesem Ensemble – Sheesh Mahal – ist heute zum Hotel umgebaut. Bekannt sind außerdem die vielen Grabstätten – Kenotaphen oder Chattris (ind.) – der Bundella-Könige entlang der Betwa, leider oftmals stark verfallend.

Agra – Taj Mahal

Das Denkmal der Liebe, ein Hohelied an die Symmetrie! Noch poetischer: „Eine Träne auf der Wange der Zeit" oder als Zitat einer Inschrift im Mausoleum „Sein Meister kann nicht dieser Welt entstammen, denn sichtbar gab ihm diesen Plan der Himmel".

Der Enkel Akbars, Prinz Khurram verliebte sich in die Prinzessin Arjumand Begum und heiratete sie 1612 als seine Zweit-Frau. 1628 bestieg Khurram gewaltsam den Thron (er ließ seine Brüder ermorden) und nannte sich Shah Jajan. Seine Lieblingsfrau Begum erhielt den Namen Mumtaz Mahal. 19 Jahre dauerte ihr Glück, sie starb im Wochenbett bei der Geburt ihres 14. Kindes. Shah Jahan plante ein Denkmal zur „ewigen" Erinnerung, 1631 wurde mit dem Bau begonnen. Die Aussagen zur Bauzeit differieren zwischen 11 und 22 Jahren – egal, das Ergebnis ist überwältigend!!

Der Bauplan ist eine architektonische Vereinigung vom Diesseits und Jenseits nach islamischer Vorstellung. Im Norden das Grabmal mit dem Grabgarten, dann als Übergang der Vorhof mit dem Haupttor als Eingang zum Garten und weiteren drei Eingängen. Im Süden dann der weltliche Teil mit Basaren und Karawansereien. Die Farbgebung ist ein weiterer Mosaikstein in diesem ausgeklügelten Ensem-

Land & Leute | The country and the people

Innenhof des „Jahangir Mahal", Orcha.

Courtyard at Jahangir Mahal, Orcha.

son Salim as crown prince order the expedition to conquer Orchha in 1594, he also became friends with the ruler of Orchha, Vir Singh Deo, who made further additions to the Fort of Orchha. The new palace was called Jahangir Mahal in remembrance, as Jahangir was Salim's imperial name. Jahangir Mahal is a symmetrical construction and its many elegant and delicate windows – influenced by Mughal architecture – are particularly impressive.

The Fort of Orchha stands on an island connected to the mainland by a bridge. It is often described as a palace complex because the buildings that make up the complex were not only used for defensive purposes. The royal palace, the Raj Mahal, has many impressive ceiling frescos. The third palace in the trio, the Shish Mahal, has been converted and is currently a hotel. The many tombs – cenotaphs, or chhatris (Hindi) – of the Bundela Kings along the Betwa river are also famous, although in many cases in a state of considerable decay.

Agra – Taj Mahal

The monument to love, a paragon of symmetry! Expressed even more poetically: „A tear on the cheek of time", or, to paraphrase an inscription in the mausoleum: „Its master cannot be of this world, for he was clearly given the plan by heaven".

Akbar's grandson, Prince Khurram, fell in love with Princess Arjumand Begum and took her as his second wife in 1612. In 1628, Khurram seized the throne by force (by having his brother murdered) and called himself Shah Jahan. His favourite wife, Begum, was given the name Mumtaz Mahal. They spent a happy 19 years together until she died giving birth to their 14th child. Shah Jahan planned a monument in „eternal" remembrance and construction began in 1631. Reports on the duration of construction vary from 11 to 22 years, but, however long it took, the result is breathtaking.

The design is an architectural amalgamation of this life and the afterlife according to Islamic belief. The tomb is situated to the north, with the garden directly adjacent. A forecourt leads to the main gateway as the entrance to the tomb garden, along with three further entrances. To the south is the worldlier realm of bazaars and caravanserais. The colour scheme is another tile in the mosaic of this sophisticated ensemble: Gleaming white marble for the mausoleum, while the walls of the tomb garden and the buildings of the forecourt are clad in red sandstone.

Die Gartenseite des Haupttores – von hier ist der erste Blick auf das „Taj Mahal" möglich, Agra.

The main entrance seen from the gardens – the Taj Mahal can first be seen from here, Agra.

ble. Strahlend weißer Marmor für das Mausoleum, während die Mauern des Grabgartens und die Gebäude des Vorhofes mit rotem Sandstein verkleidet sind.

Der Garten des Taj Mahal stellt ein irdisches Abbild des koranischen Paradieses dar. Die Grundfläche wird von zwei Hauptwegen in vier Quadrate geteilt und die Kanäle in den Hauptwegen sind paradiesischen Flüssen nachempfunden. Im Zentrum des Gartens münden die Kanäle in ein Wasserbecken, ein Sinnbild des himmlischen Brunnens, an dem die Gläubigen ihren Durst stillen, wenn sie ins Paradies eintreten.

Der Innenraum des Mausoleums ist nach dem Vorbild der acht Paradiese des Korans gestaltet – acht Kammern umgeben den zentralen Hauptraum unter der Kuppel. Hier stehen die Kenotaphe von Mumtaz Mahal und von Shah Jahan. Der eigentliche Sarg von Mumtaz Mahal steht in einer Marmorkammer unter dem Hauptraum, niemand soll die Ruhe der Toten stören!

Einige Fakten/Zahlen zum Bau:
- 1.000 Elefanten schleppten die Marmor- und Sandsteinplatten zur Verkleidung der Mauern, die aus vor Ort gebrannten Ziegeln bestehen.
- Die Plattform für das Mausoleum, Moschee sowie Gästehaus ruhen auf einem Gerüst, das unterhalb der Grundwasserlinie beginnend aufgebaut wurde.
- 20.000 Arbeiter waren im Einsatz, um beispielsweise die schweren Marmorplatten für die Kuppel über eine 3 km lange Rampe nach oben zu transportieren.
- Der Komplex mit Garten und Mausoleum misst etwa 600 x 600 m und ist von einer Sandsteinmauer umgeben.
- Den Eingang bildet das etwa 30 m hohe Haupttor, der Sammelpunkt des Vorhofes. Die Ausführung gleicht der des Siegestores in Fatehpur Sikri.
- Das Hauptgebäude hat einen quadratischen Grundriss von etwa 57 m Länge und steht auf einer 7,5 m hohen und ebenfalls quadratischen Plattform.

Land & Leute | The country and the people

The garden of the Taj Mahal represents an earthly portrayal of paradise as described in the Koran. The main area is divided into four quarters by two main pathways, and the canals running along these pathways are meant to symbolise the four rivers of paradise. In the centre of the garden the canals empty into a water tank in reference to the Tank of Abundance at which the faithful quench their thirst when they enter paradise.

The interior of the mausoleum is modelled on the eight paradises of the Koran – eight chambers surround the inner chamber under the dome. The inner chamber houses the cenotaphs of Mumtaz Mahal and Shah Jahan. Mumtaz Mahal's actual coffin is in a marble crypt beneath the inner chamber, allowing the deceased to rest in peace.

A few facts and figures on the construction:
- 1,000 elephants were needed to drag the marble and sandstone slabs to clad the walls, which are built from bricks fired locally.
- The platform for the mausoleum, the mosque and the guest house are on a structure whose foundations begin below the water line.
- 20,000 workers were required for the construction, with some needed for transporting the heavy marble slabs up a 2-mile-long ramp to the dome.
- The complex, including garden and mausoleum, measures around 2,000 x 2,000 ft and is surrounded by a sandstone wall.
- Entry is via the 100-ft-high Great Gate, the forecourt meeting point. It is decorated in the same style as the Victory Gate at Fatehpur Sikri.
- The main building has a square floor plan of around 185 ft on each side and stands on a square platform that is 25 ft high.
- Standing 185 ft high, the central building is essentially a cube. This impression is softened by the slender minarets at the corners of the platform, which are each approximately 140 ft high.

Holi celebrations
Holi, the festival of colours, has spread beyond India and is now celebrated in parts of Europe, too, including Germany

Das „Taj Mahal", vom Garten aus betrachtet.

The Taj Mahal as seen from the garden.

- Der Zentralbau ist mit 57 m Höhe fast würfelförmig, gemildert wird dieser Eindruck von den schlanken Minaretten an den Ecken der Plattform, jeweils ca. 42 m hoch.

Holi-Fest

Das Holi-Fest, das Fest der Farben, wird heute vielfach auch außerhalb Indiens – sogar in Deutschland – gefeiert. Es ist in unterschiedlichen Ausprägungen in ganz Indien verbreitet und dauert meistens zwei Tage. Fünf Tage nach Vollmond werden Elefanten geschmückt, bemalt und in einem Festzug vorgeführt. Über alle Gesellschaftsschichten, Kasten, Geschlechter oder Alter hinweg wird farbiges Wasser verspritzt bzw. Puderfarben verstreut oder aufgetragen, oder es werden Blütenblätter verstreut. Am Tag davor werden meistens Feuer entzündet und darin (Holi-)Puppen verbrannt. Die Legende erzählt: *Der König Kashyap befahl seinen Untertanen, ihn als Gott anzubeten. Nur sein Sohn Prahlad verehrte weiterhin Vishnu. Kashyap versuchte verschiedentlich seinen Sohn zu töten, doch dieser überlebte alle Anschläge. Da hatte Kashyap die Idee, dass seine Schwester Holika – diese besaß einen durch Zauber feuerfesten Schal – mit Prahlad auf dem Schoß auf einen Scheiterhaufen steigen sollte. Holika willigte ein, Vishnu sandte jedoch einen Windstoss, der den Schal von Holika löste und um Prahlad wickelte, sodass dieser überlebte und Holika verbrannte.*

Das farbenfroheste Fest soll in Mathura, der Stadt Vishnus, nahe Agra, gefeiert werden. Wir erlebten in Kahora eine ganz spezielle Form: Schulkinder bewarfen und beschmierten sich mit Schlamm – alles ist auch ein Symbol für den beginnenden Frühling, für den Triumph des Guten über das Böse und für den Versöhnungsaspekt – alle Streitereien sollen beendet werden. Unsere eigene Bemalung mit Farben wurde bei unserem Besuch in einem Hundertseelendorf versteckt aber intensiv bewundert, ganz mutige Mädels machten mit ihren Handys sogar Bilder von uns!

Alltagsleben

Außerhalb der Metropolen dominieren Ackerbau und Viehzucht sowie Handel die Erwerbstätigkeit – Indien versorgt sich mit den Grundnahrungsmitteln weitgehend selbst. Supermärkte und mehrstöckige Einkaufszentren werden in den kleineren Städten durch Straßen ersetzt, an denen sich kleine Läden und Verkaufsstände wie an einer Perlenschnur aneinanderreihen. In den meisten wartet ein Mann auf Kundschaft und verkürzt sich die Zeit mit einem Gläschen Tee und Schwätzchen mit seinen Kumpels. In den Dörfern kann man auch noch die Herstellung von Feuermaterial, getrocknetem Kuhmist, bewundern und natürlich die fröhlich-freundliche Neugier der Kinder und vielfach auch der Mütter erleben.

Ein typischer Marktstand mit eigenen Produkten.

A typical market stall with home-made produce.

Land & Leute | The country and the people

Bemalung zur Feier des Holi-Festes.

Face-painting in celebration of Holi.

and North America. Holi takes various forms throughout India and usually lasts for two days. Five days after full moon, elephants are decorated, painted and led in a procession. All layers of society and all castes join in. People of all ages and both sexes throw or apply coloured water and powders or petals. A Holika bonfire is usually lit the night before Holi and effigies are burned. Legend has it that *King Kashyap ordered his subjects to worship him as a God. Only his son Prahlad continued to worship Vishnu. Kashyap made various efforts to kill his son, but he survived them all. Kashyap then had the idea of having his sister Holika – who had a magic cloak that prevented her from being harmed by fire – climb up and sit upon a pyre with Prahlad on her lap. Holika agreed, but Vishnu sent a gust of wind that blew the cloak from Holika and wrapped it round Prahlad, allowing him to survive while Holika burnt to death.*

This colourful festival is thus inseparable from Vishnu's home city of Mathura, near Agra. We witnessed a very special form of the celebration in Kahora: Schoolchildren threw mud and smeared it on each other, all as a symbol of the onset of spring, the triumph of good over evil and of reconciliation – with all disputes to be brought to an end. The way we were covered in bright colours was the subject of hidden but close scrutiny and admiration in a small village of a hundred or so inhabitants; some particularly courageous young girls even took photos of us on their mobile phones!

Everyday life

Outside the cities, Indians primarily earn their livelihood with agriculture, livestock farming and trade – India is largely self-sufficient in producing its staple foods. Smaller towns have streets full of little shops and stalls lined up like pearls on a string rather than supermarkets and multi-storey shopping centres. Most of them are attended by a man waiting for customers and passing the time with a glass of tea and chatting with friends. In villages, tourists can watch the production of solid fuel – dried cow dung – and enjoy being the subject of the friendly curiosity of the children and, often, of their mothers, too.

Autor/Author: Harald Lydorf

Papier-Transport in „Old Town Delhi" per Muskelkraft.

Paper is transported using traditional methods in Delhi old town.

Auch die Nutztiere werden geschmückt (oben), eine Näherin in ihrem Haus (unten).

Even working animals are decorated (top); a seamstress at home (bottom).

Bambusstangen werden für das Flechten von Matten zerfasert (oben), Teepflückerinnen kommen von der Arbeit (Mitte), eine Opfergabe im Tempel (unten).

Bamboo shoots are prepared for weaving into mats (top); tea-pickers return from the fields (middle); a temple offering (bottom).

Die kulturelle Vielfalt Indiens auf einer Seite: Der Blick vom „Jajangir Mahal" in Orcha auf seine Nebengebäude und eine Grabstätte (oben), der „Pachisi Platz" mit "Diwan-I-Kas" in Fatehpur Sikri (Mitte), der „Chitragupta Tempel" in Khajuraho (unten links), der 70 Meter hohe Turm in Qutub Minar in Delhi (unten rechts) und ein Bogengang im Vorhof des „Taj Mahal", Agra (rechte Seite).

India's diverse culture all on one page: The view of neighbouring buildings and a burial site from Jajangir Mahal in Orchha (top); the Pachisi court and Diwan-I-Kas in Fatehpur Sikri (middle); Chitragupta Temple in Khajuraho (bottom left); the Qutub Minar tower in Delhi, which stands 230 ft tall (bottom right); a series of doorways on the grounds of the Taj Mahal, Agra (right).

Detail der Steinmetzarbeiten im Marmor des „Taj Mahal" (oben links), erotische Szene am „Lakshmana-Tempel" in Khajuraho (oben rechts).

Intricate carvings in the marble of the Taj Mahal (top left); erotic depictions from the Lakshmana Temple in Khajuraho (top right).

Durchblick zum Turm von „Qutub Minar" in Delhi (unten links), ein Säulengang im „Roten Fort" in Agra (unten rechts).

View to the Qutub Minar tower in Delhi (bottom left); a gallery of pillars at Agra Fort (bottom right).

Roter Sandstein des Gästehauses im Gegensatz zum weißen Marmor des „Taj Mahal" (oben links), der Kandariya Mahadeo Tempel in Khajuraho (oben rechts).

The red sandstone of the guesthouse contrasts with the white marble of the Taj Mahal (top left); the Kandariya Mahadeo temple in Khajuraho (top right).

Reich verzierte Stützbalken im Roten Fort in Agra (unten links). Ornamente im weißen Marmor des „Taj Mahal" (unten rechts).

Heavily embellished supporting beams at Agra Fort (bottom left). Decorations in the white marble of the Taj Mahal (bottom right).

Vielfältigste Eindrücke der indischen Lebensart lauern überall: Die mühselige Ziegelherstellung erfolgt Tag für Tag gleich in der Nachbarschaft einer Fünf-Sterne-Touristen-Lodge (oben), den Heimweg der Arbeiterinnen nach der Arbeit auf den Reisfeldern sahen wir am späten Nachmittag vom Auto aus (unten).

The Indian way of life is full of contrasts: Every day, bricks are produced laboriously using manual methods while, nearby, tourists stay in a five-star lodge (top); paddy field workers leave for the day, as seen from the car in the late afternoon (bottom).

Ranthambore

Der Tiger ist wieder Herr in seinem Land

Ranthambore

The tiger returns to dominance

Ranthambore

Der Ranthambore-Nationalpark ist eine der attraktivsten und bekanntesten Naturattraktionen im indischen Bundesstaat Rajasthan. Überragt wird er von den Ruinen des 944 begonnenen Festungsbaus „Ranthambore", der dem Park seinen Namen gab. Nach wechselnden Besitzern unter den Hindu-Königreichen wurde das Fort 1559 vom Mogulkaiser Akbar erobert. Später kam es in den Besitz der Maharadschas von Jaipur und das umliegende Areal wurde zu deren Jagdgebiet – und blieb dies bis zum Jahre 1970 – also selbst noch nach der Einrichtung des ersten Schutzgebietes!

Bereits 1955 richtete die indische Verwaltung das etwa 400 km² große „Sawai Madhopur Wildschutzgebiet" ein und machte dieses 1973 zum Teil des landesweiten Projektes „Schutz der Tiger". Aus dem Kerngebiet entstand 1980 der etwa 280 km² große „Ranthambore-Nationalpark". 1984 wurden im Norden und Süden zwei Waldgebiete mit zusammen rund 900 km² zu Wildschutzgebieten erklärt, die nun zusammen mit dem Nationalpark das „Ranthambore Tiger Reservat" bilden.

Für uns ging es Schlag auf Schlag: Flug von Frankfurt nach Delhi; Transfer ins Hotel mit einem ersten Eindruck der indischen Verkehrsdichte inklusive Nachhilfe im intensiven Gebrauch der Hupe; kurze Nachtruhe; Fahrt zum Bahnhof; danach noch eine 3-stündige Zugreise im „Goldenen Tempel" (der Name des Zuges) nach Sawai Madhopur, ca. 160 Kilometer südöstlich von Jaipur. Sawai Madhopur ist der Ausgangspunkt für einen Besuch im Ranthambore Nationalpark. Hier ist der Sitz der Parkverwaltung und hier gibt es sehr viele Hotels, man erkennt die Bedeutung des Tourismus – mit dem Tiger als Anziehungsmagnet – an jeder Ecke.

Die Felder der Bauern grenzen direkt an die zum Nationalpark gehörenden Berghänge.

The farmers' fields directly adjoin the mountain slopes of the national park.

Ranthambore

Ranthambore National Park is one of the most beautiful and well-known natural attractions in the Indian state of Rajasthan. It is overlooked by the ruins of Ranthambore Fort, construction of which began in 944 and which gave the park its name. The fort changed hands many times throughout the various Hindu empires, before being conquered by Akbar, the Mughal Emperor, in 1559. Later still, the fort passed into the hands of the Maharajas of Jaipur, who turned it and the surrounding area into a royal hunting reserve, which it remained until 1970 – even after the designation of the first protected area.

Back in 1955, the Indian government created the Sawai Madhopur Game Sanctuary covering an area of about 150 square miles, before, in 1973, declaring it one of the reserves belonging to the national Project Tiger initiative. In 1980, Ranthambore National Park was created from core section of the park, some 110 square miles in size. In 1984, two forested areas to the north and south, covering a total area of some 350 square miles, were also turned into wildlife sanctuaries. Together with the National Park, these now form the Ranthambore Tiger Reserve.

For us, the journey barely left time to pause for thought. We flew from Frankfurt to Delhi, and the hotel transfer quickly gave us an idea of just how heavy the traffic is on Indian roads, accompanied by much blowing of horns. A short night was followed by a trip to the station and a three-hour journey with a train called the „Golden Temple" to Sawai Madhopur, around 100 miles south-east of Jaipur. Sawai Madhopur is the starting point when visiting Ranthambore National Park. The park authorities are based there and the place is packed with hotels, showing how important tourism is for the area – and it is

Ranthambore National Park	
State	Rajasthan
Nearest city	Sawai Madhopur
Size [sq mi]	108 (core zone)
Protected since	1955
National Park since	1980
Best time to travel	November – May
Zones	10
Landscape	Bush, mixed deciduous forest, grassland, rugged cliffs, lakes
Fauna	
Predators	Bengal tiger, leopard, sloth bear, jungle cat, mongoose
Grazing animals	Nilgai (blue bull antelope), sambar deer, chital (spotted deer), chinkara (Indian gazelle), wild boar
Birds	Peafowl, plum-headed parakeet, rose-ringed parakeet, yellow-wattled lapwing, Indian roller, painted stork, bay-backed shrike, white-throated kingfisher
Primates	Grey (hanuman) langur, rhesus macaque
Other	Mugger crocodile

Selbst die Ruinen lassen die Pracht der einstigen Jagdpaläste erahnen.

Even the ruins hint at the magnificence of the former hunting palaces.

Ranthambore

Ranthambore Nationalpark	
Bundesstaat	Rajasthan
nächstgelegene Stadt	Sawai Madhopur
Größe [km²]	280 (Kerngebiet)
Schutzgebiet seit	1955
Nationalpark-status seit	1980
beste Reisezeit	November – Mai
Zoneneinteilung	10
Landschaft	Buschland, Laubmischwald, Grasland, schroffe Felsen, Seen
Tierwelt	
Raubtiere	Bengaltiger, Leopard, Lippenbär, Rohrkatze, Mungo
Grasfresser	Nilgauantilope, Sambarhirsch, Axishirsch, Indische Gazelle, Wildschwein
Vögel	Pfau, Pflaumenkopfsittich, Halsbandsittich, Gelblappenkiebitz, Hinduracke, Buntstorch, Rotschulterwürger, Braunkopfliest
Primaten	Hanuman-Langur, Rhesusaffe
Andere	Sumpfkrokodil

Das Eingangstor zu den Zonen 1 bis 5.

The entrance to zones 1 to 5.

Deutlich zu sehen ist aber auch, welche Belastungen für die Umwelt der ständig steigende Gebrauch von Plastikverpackungen mit sich bringt – überall liegt Müll herum und verrottet nicht, eine Müllabfuhr gibt es nicht ...

Ziemlich genau 25 Stunden nach unserem Start in Deutschland wurden wir von zwei Guides, die uns die nächsten Tage begleiteten, und zwei Fahrern mit ihren Jeeps abgeholt. Die Fahrer wechseln bei jeder Safari. Drei Stunden hatten wir nun die Chance und die Hoffnung, Tiger zu sehen und zu fotografieren! Tigersafaris konzentrieren sich auf das Kerngebiet und können nur bei der Parkverwaltung gebucht werden. Durchgeführt werden sie zwei Mal am Tag (früher Vormittag und später Nachmittag) mit je drei Stunden Dauer. Als Fahrzeuge stehen Canter, Lkw mit 25 Sitzen, und Gypsys, kleine Jeeps mit 6 Sitzen, zur Verfügung. Verkauft werden einzelne Sitzplätze, dazu kommen noch zusätzliche Gebühren, wie beispielsweise für Kameras. Es geht aber nicht dergestalt los, dass nun der Guide aus seiner Erfahrung einen Kurs vorschlägt – nein, er holt sich vor der Abfahrt eine Zuteilung für eine „Zone" bei der Parkverwaltung ab. Darin gibt es eine vorgegebene Strecke und die Wege dürfen nicht verlassen werden. Ranthambore ist in zehn Zonen eingeteilt und diese werden per Los (jedenfalls offiziell) auf die für den jeweiligen Termin gebuchten Fahrzeuge verteilt. Hintergrund ist der Versuch, einerseits möglichst vielen Besuchern den Einlass zu ermöglichen, andererseits aber den Stress der Tiger in den Hotspots durch eine zu hohe Dichte von Fahrzeugen zu mindern.

Am Eingang des Parks wurden regelmäßig zwei Kämpfe ausgetragen: Erstens unser Guide mit der Parkverwaltung (man muss als Besucher übrigens immer seinen Pass dabei haben, um sich als die im Permit aufgeführte Person ausweisen zu können) bis zum Erhalt des Stempels –

the tigers that draw the crowds. The piles of rubbish lying around clearly reveal the impact of the rising consumption of plastic packaging on the environment, and with no refuse collection here it simply gets left and doesn't biodegrade ...

Almost exactly 25 hours after setting off in Germany, we were met by two guides who would show us around over the next few days, along with two drivers and their vehicles. We had different drivers on each safari. We set off with three hours ahead of us in which we hoped to see and take pictures of some tigers. Tiger safaris stay in the core area of the park and can only be booked via the park authorities. There are two per day (early morning and late afternoon), each lasting three hours, and can be taken in 25-seater Canter trucks or six-seater Gypsy off-road vehicles. Seats are sold individually, with extra charges such as for cameras. But you needn't think the route of our safari is determined by what the guide suggests, based on his experience. No; he is assigned a specific zone by the park authorities before setting off. He is given a permit with a predetermined route, which we were not allowed to deviate from. Ranthambore is split into 10 zones, which are randomly assigned to the vehicles booked for the day by drawing lots (that's the official version, anyway). This is because they want to try and let as many people in as possible while keeping to a minimum the level of stress for the tigers caused by the large number of vehicles at the hotspots.

There were usually two battles going on at the park entrance. First, our guide had to

Affenliebe, eine Languren-Mutter behütet ihren Nachwuchs.

A female langur protects her young.

T24, Ustad – ein prachtvoller männlicher Tiger.

T24, Ustad – a magnificent male tiger.

Ranthambore

und zweitens wir mit den fliegenden Händlern, die sehr hartnäckig auftraten! Schließlich rumpelte unser kleiner Jeep mit seiner schwachen Federung, dem kurzen Radstand und den schmalen, nahezu ungepolsterten Sitzreihen über steinige, staubige Straßen in die zugewiesene Zone. Landschaftlich durchaus reizvoll kamen bald die ersten Bewohner ins Zielfeld unserer Kameras: Axis- und Sambarhirsche, verschiedene Vogelarten und Affen.

Fast schon romantisch wirkten die verlassenen Jagdpaläste an einem der Seen, aber wir mussten uns bei fantastischem Licht bereits wieder auf den Rückweg machen – das Tor schließt pünktlich und Fahrer und Guide drohen bei Zeitüberschreitung deftige Strafen, einschließlich zeitweisem Verbot zum Betretens des Parks! Dann hatten wir aber eine riesige Portion Glück: Knapp einen Kilometer vor dem Ausgang und 10 Minuten vor Schluss kam unser erster Tiger, T25, einen Abhang runter, legte sich in einen kleinen, algenbewachsenen Tümpel neben der Straße, trank und verschwand wieder im Dickicht – das ganze Schauspiel hatte rund 5 Minuten gedauert und uns einige schöne Bilder beschert, sowie unseren Adrenalinspiegel mächtig in die Höhe getrieben.

Bei manchen unserer Safaris war die Ausbeute nicht so toll, aber insgesamt konnten wir als Team sehr schöne Bilder machen und viele Eindrücke – auch von den Menschen der angrenzenden Dörfer – mitnehmen. Bei einem unserer wiederholten Besuche in der Zone 6, hier sollten sich Lippenbären bevorzugt aufhalten, wurden wir – und unsere Kameras! – Zeuge einer seltenen Jagd. Ein Braunkopfliest saß in einem Busch am Rande eines trockenen Flussbettes, so dachten wir. Was wir nicht sahen, war ein übrig gebliebener kleiner Tümpel mit Bewohner – einer Süßwasserkrabbe! Die wurde für uns erst sichtbar, als sie bereits im Schnabel des Liest zappelte.

In dieser Zone konnten wir auch erleben, wie sehr die Mitarbeiter der Parkverwaltung an eigenen Tigersichtungen interessiert sind: Bei einer Ausfahrt, nahe am Gate – schon spät und bei schlechtem Licht – kreuzte ein Tiger den Weg. Wir „mussten" mit den drei Rangern vom Gate nochmals zurückfahren, hatten dann aber kein weiteres Glück und die Enttäuschung der Ranger war groß.

Eine Belohnung für unsere vielen Versuche gab es in Zone 1: Unser Guide hatte über „Buschfunk" die Info bekommen, dass T72, Sultan, ein jüngerer männlicher

Reste eines kleinen Flusses, sonst dominiert die Farbe der Tiger.

The last vestiges of a small river, dominated otherwise by the colour of the tigers.

Ranthambore

Ein Leopard auf dem höchsten möglichen Aussichtspunkt.

A leopard observes from up on high.

Idyllisch: Vollmond über Ranthambore.

Idyllic: Full moon over Ranthambore.

get the park authorities to stamp the permit (visitors have to carry their passports at all times, by the way, to prove they are the person named on the permit), and, second, we had to make our way through the street vendors, who could be very persistent. Eventually, our small off-road vehicle with worn out suspension, a short wheelbase and narrow, virtually unpadded benches trundled its way via stony, dusty roads to our allocated zone. Scanning the beautiful landscape, the first chital (spotted deer) and sambar deer moved into our viewfinders, along with several species of birds and simians.

The deserted hunting palaces by one of the lakes had an almost romantic feel to them, but despite the fantastic light we had to head back. The gates close punctually and the drivers and guides can have hefty fines imposed if they are late, including being temporarily barred from the park. We were really lucky on the way back, however – less than 1,000 yards from the exit and 10 minutes before closing time, we saw our first tiger. Known as T25, it came down a slope, lay down in a small algae-covered pond right next to the road, had a drink and disappeared back into the bushes. The whole thing lasted around five minutes and gave us some great pictures, not to mention sending our adrenalin levels through the roof.

We weren't as lucky on all of our safaris, but the team as a whole got some really good pictures and lots of memories, including of the people in the nearby villages. During one of several safaris in zone 6, which we were told was a hotspot for sloth bears, we – and our cameras, of course – witnessed a curious hunt. A white-throated kingfisher was sat in a bush at the edge of a dry river bed. Or so we thought. What we couldn't see were the last vestiges of a small pond, home to a freshwater crab. We only saw the crab once it was already floundering in the kingfisher's beak.

It was in this zone that we also experienced how eager the park employees were to see the tigers for themselves. As we were leaving one day – it was already late and the light was fading – a tiger crossed

Ranthambore

Tiger gesichtet wurde. Nach langer Fahrt durch ein bewaldetes Tal kamen wir an einer Furt an und sahen zunächst nur kreuz und quer stehende Jeeps. Für uns blieb nach Querung der Furt nur ein Platz in der zweiten Reihe: Wir sahen Sultan kaum durch die Zweige, halb im Gegenlicht liegend, aber eindrucksvoll gähnend. Das wollten wir besser haben und überredeten unseren Guide, etwa 100 Meter entfernt an das andere Ende des Tümpels zu fahren und dort zu warten. Unsere Hoffnung war, dass der Tiger sich später am Nachmittag in diese Richtung bewegen würde, wo wir dann freie Sicht und auch noch gutes Licht hätten ... er tat uns den Gefallen – Glück gehabt!

T25, Dollar oder Zalim

T25, farbig bestens vorbereitet für das Holi-Fest – ein wirklich farbenfrohes Frühlingsfest *(mehr dazu im Teil „Land und Leute")* – zeigt sich nicht nur hier als exzentrischer Charakter! Als Halbstarker machte sich dieser jetzt etwa 6 Jahre alte männliche Tiger wohl zeitweise einen Spaß daraus, die Jeeps der Touristen zu jagen! Zudem trägt er nicht nur eine nüchterne Zahl, sondern schmückt sich gleich noch mit zwei Namen: „Zalim", was soviel wie der „Grausame" heißt und „Dollar", weil die Streifen auf seiner rechten Flanke mit etwas Fantasie ein Dollarzeichen bilden.

Groß war dann im Februar 2011 die Überraschung, als T25 die beiden weiblichen Waisen „Bina 1 und 2" adoptierte! Er beschützte sie nicht nur, sondern trainierte sie auch wie eine Mutter – völlig gegensätzlich zum sonstigen Verhalten männlicher Tiger, die sich nicht um die Aufzucht der Jungen kümmern.

Das nächste Mal verblüffte er die Beobachter, als er sich mit T17, „Sundari" oder auch „Satara", der „Königin der Seen-Region" paarte und diese, nach früheren Fehlversuchen mit anderen männlichen Tigern, im Mai 2012 drei Junge gebar!

Vielleicht hat er in Zukunft noch andere Überraschungen auf Lager ...

Der Wanderbaumelster wird der Hirschkuh lästig (oben), ein blauer Pfau im Porträt (unten).

A rufous treepie irritates a deer (top); an Indian peafowl in profile (bottom).

the track just near the gates. We simply had to go back with the three rangers from the gate, but the tiger had already gone and the rangers were extremely disappointed.

We were rewarded for the number of trips we made during one safari in zone 1 where our guide had heard via the „bush telegraph" that a young male tiger called T72, also known as Sultan, had been sighted. After a long drive through a wooded valley, we arrived at a ford and the first thing we saw were safari vehicles parked all over the place. After crossing the ford, there wasn't much room left for us and through the branches we could hardly see Sultan as he had the sun behind him. We did catch some impressive yawns, though. But we thought we could do better and convinced our guide to drive up to the other end of the pond, around 100 yards away. We waited and hoped that Sultan would come along this way later in the afternoon as we had a great view and the light was good ... and, as luck would have it, he did just that!

T25, Dollar or Zalim

T25's superb colouring makes him look like he has been created especially for the colourful Holi celebrations in spring (read more in the chapter „The country and the people") and his eccentric character comes through everywhere. As a teenager this now around six years old male tiger seemed to enjoy hunting the tourists in their vehicles. He's not known only by a mundane number; he also has two names – Zalim, which translates roughly as „the ferocious one", and Dollar, because with a little imagination the stripes on his right side could look like a dollar sign.

So it was to everyone's surprise when, in February 2011, T25 adopted two female orphans, Bina 1 and Bina 2. He protects them and teaches them like a mother – which is completely at odds with the normal behaviour of male tigers, who have nothing to do with rearing their young.

Another surprise was in store when he mated with T17, known as Sundari or Satara, „the queen of the lakeland region", who gave birth to three cubs in May 2012 after failed attempts with other males. Perhaps there are even more surprises yet to come ...

Autor/Author: Harald Lydorf

T25, Dollar, ein männlicher Tiger mit grünem „Holi"-Farbschmuck nach einem Bad in einem Algentümpel.

Male tiger T25, or Dollar, wears green for the Holi after bathing in an algae-covered pond.

Nilgauantilope, junger Bulle (oben), Feigen verdecken mit ihren Wurzeln ein Tor fast komplett (unten).

A young male nilgai (top); an entrance is almost concealed by fig tree roots (bottom).

Im Uhrzeigersinn von unten:
Reiher, Buntstorch und
Wellenbrust-Fischuhu.

Clockwise, from bottom left: Egret,
painted stork and brown fish owl.

Ein höchst seltener Anblick: Der Lippenbär als Straßengänger.

Sloth bears are rarely seen on the roads.

Ein Reiher im Ansitz.

An egret perches patiently.

Auch wenn Tigerbegegnungen keinesfalls garantiert sind ... während nur weniger Tage Aufenthalt im Ranthambore Nationalpark haben sich uns doch mehrere und sehr unterschiedliche Foto-Gelegenheiten geboten.

Despite there being no guarantee of seeing tigers... we managed to capture plenty of pictures over various photo opportunities during our few days in Ranthambore National Park.

So etwas sieht man nur äußerst selten. Diese zwei Tiger kämpfen heftigst miteinander. Da treffen echte Naturgewalten aufeinander ...

An extremely rare sight. Two tigers fiercely battle it out. These animals are real forces of nature...

Viel Zeit verbringen Tiger während des Tages mit Ausruhen und Trinken.

Tigers spend a lot of the day relaxing and drinking.

Ein Braunkopfliest verspeist eine Süßwasserkrabbe (rechts), Schlangen bei der Paarung (links), Sumpfkrokodil beim Sonnenbad (unten).

A white-throated kingfisher eats a freshwater crab (right); snakes mating (left); a mugger crocodile basks in the sun (bottom).

Kaziranga

Weltnaturerbe mit über 100 Jahren Tradition

Kaziranga

World Heritage Site with a history that goes back more than 100 years

Der Kaziranga-Nationalpark liegt im Bundesstaat Assam – Teekenner schnalzen jetzt mit der Zunge – im äußersten Nordosten des Landes, einer von Nepal, China, Myanmar und Bangladesch umgebenen Region Indiens. Kaziranga ist seit neuestem etwa 860 km² groß und liegt beiderseits des Brahmaputra, welcher das Gebiet und das Leben dort entscheidend prägt. Südlich vom Fluss bis zur Nationalstraße 37 liegen vier Sektoren mit rund 400 km² Fläche. Diese bilden den allgemein bekannten und von Touristen besuchbaren Teil des Nationalparks.

Im Jahr 2005 feierte der Park sein hundertjähriges Bestehen! Vom Waldreservat über ein Schutzgebiet für Panzernashörner 1916 und Nationalpark seit 1974, zum Weltnaturerbe 1985 und schließlich auch noch Tigerreservat seit 2007. Kaziranga beherbergt etwa 2.000 Panzernashörner und somit rund 70 % des weltweiten Bestandes, 1.800 wilde Wasserbüffel, 1.300 wilde Indische Elefanten, 9.000 Schweinshirsche, 800 Zackenhirsche (Barasingha) und 90 bis 100 Bengaltiger, die höchste Tigerdichte Indiens, bezogen auf die Fläche. Dazu kommen neben anderen Raubtieren wie Leoparden und Schleichkatzen noch Warane und Schlangen und eine Vielzahl von Vogelarten.

Unsere Nacht im Hotel in Delhi war kurz, die Fahrt zum Flughafen auch, und das Einchecken lief wie geschmiert. Das verlockte uns zu einem lange entbehrten Espresso, was sich fast rächte – der Weg durch die Sicherheitskontrollen dauerte ewig, sodass wir gerade noch mit dem

Ein Panzernashorn-Bulle versucht Kontakt mit einer Kuh aufzunehmen, die aber noch ein Kalb führt.

A male Indian rhino attempts to attract the attention of a female, but she is still with her calf.

Kaziranga

Kaziranga National Park is in the state of Assam – a tea lovers' paradise – in the far north-east of the country, with Nepal, China, Myanmar and Bangladesh virtually a stone's throw away. Recently extended to cover an area of around 330 square miles, Kaziranga sits on both sides of the Brahmaputra River, which is a vital lifeline for the region. The area stretching south of the river to National Highway 37 is split into four sectors covering some 150 square miles. These sectors are accessible to tourists and are generally the best known part of the national park.

Kaziranga National Park	
State	Assam
Nearest city	Jorhat
Size [sq mi]	166 (core zone)
Protected since	1905
National Park since	1974
Best time to travel	November – April
Zones	4
Landscape	Elephant grass, grassland, jungle, waterways, swampland, woodland
Fauna	
Predators	Bengal tiger, Asian black bear, viverrid
Grazing animals	Asian elephant, Indian rhinoceros, water buffalo, sambar deer, barasingha (swamp deer), wild boar
Birds	Vulture, crested serpent eagle, rose-ringed parakeet, pelican, Asian openbill stork, ruddy shelduck, bee eater
Primates	Rhesus macaque
Other	Bengal monitor, rat snake, cobra

In 2005, the park celebrated its centennial. Starting life as a protected forest reserve, it was designated a protected area for the Indian rhinoceros, or greater one-horned rhinoceros, in 1916, before becoming a National Park in 1974. It gained World Heritage status in 1985, and, finally, became a tiger reserve in 2007. Kaziranga is home to some 2,000 Indian rhinos, which make up around 70 per cent of the global population, as well as 1,800 wild water buffalo, 1,300 wild Indian elephants, 9,000 Indian hog deer, 800 barasingha, or swamp deer, and 90 to 100 Bengal tigers – the densest population of tigers in all of India in terms of the area they inhabit. They are joined by predators, such as leopards and viverrids, as well as monitor lizards, snakes and a wide variety of birdlife.

Our night at the hotel in Delhi was short, and the transfer to the airport even shorter. Check-in went smoothly, so we decided to satisfy our craving for espresso after going without for so long. We almost regretted it, though, when it took so long to

Dieser große Bengalenwaran sonnt sich vor seinem Bau.

This large Bengal monitor basks near its burrow.

Kaziranga

letzten Bus das Flugzeug nach Guwahati erreichten! Nach zwei Stunden Flug wurden wir auf drei Autos verteilt und die Fahrt ging los in Richtung Osten. Pausen zum Essen waren zwar bei insgesamt sieben Stunden Fahrtzeit sehr wichtig, aber viel schöner und auch zeitraubender war ein Fotostopp bei einer Teeplantage während der Rückkehr der Pflückerinnen von der Arbeit. Manche der Damen ließen sich bereitwillig fotografieren und posierten sehr freundlich, andere wiederum lehnten Aufnahmen strikt ab – eine Erfahrung, die sich durch die nächsten Tage hindurchzog.

Die ersten Panzernashörner sahen wir schon beim Erreichen des Parkgebietes aus der Ferne, das stärkte unsere Zuversicht, ihnen noch näher zu kommen. Die schön gelegene Lodge in Kaziranga bot

Vom Elefantenrücken aus lassen sich die Nashörner viel leichter entdecken, aber keineswegs leicht fotografieren ...

Rhinos are much easier to spot from the back of an elephant, but not easier to photograph...

Kaziranga Nationalpark	
Bundesstaat	Assam
nächstgelegene Stadt	Jorhat
Größe [km²]	430 (Kerngebiet)
Schutzgebiet seit	1905
Nationalpark-status seit	1974
beste Reisezeit	November – April
Zoneneinteilung	4
Landschaft	Elefantengras, Grassavanne, Dschungel, Flussläufe, Sumpfland, Waldungen
Tierwelt	
Raubtiere	Bengaltiger, Kragenbär, Schleichkatze
Grasfresser	Asiatischer Elefant, Panzernashorn, Wilder Wasserbüffel, Sambarhirsch, Zackenhirsch, Wildschwein
Vögel	Geier, Schlangenweihe, Halsbandsittich, Pelikan, Silberklaffschnabel, Rostgans, Bienenfresser
Primaten	Rhesusaffe
Andere	Bengalenwaran, Rattennatter, Kobra

rustikale Unterkünfte, aber weder Bier noch eisgekühlte Cola – wir mussten uns an diesem Abend drei kühle und drei warme Flaschen Sodawasser teilen ...

Elefanten-Safari – wieder frühes Aufstehen, kein Frühstück und mit zwei Jeeps ab in den Park zum Startplatz. Wir wurden jeweils zu zweit hinter einem Mahut platziert und dann schaukelte der Elefant auch schon los und zwar heftig! Zudem ist so ein Elefantenrücken – auch ein Indischer – ganz schön breit und man sitzt nicht wirklich bequem ... aber wir waren ja auch zum Fotografieren und nicht zum Vergnügen da!

Vom Elefant aus hat man einen guten Blick über etwa anderthalb Meter hohes Elefantengras und wir sahen recht bald die ersten Panzernashörner. Diese waren aber ziemlich versteckt und auch die Perspektive von so hoch oben war nicht vorteilhaft für diese Dickschiffe, die wie aus dem Jurassic-Park entsprungen wirkten. Diese Gedanken wurden abrupt gestört, als etwa 30 m entfernt ein Fauchen ertönte – ein Tiger mit einem mächtigen

get through security that we only just made the last bus for the plane to Guwahati. After two hours in the air, our group was met by three cars and taken east, making sure to stop for food on our seven-hour journey. Much more importantly, though, we also stopped at a tea plantation to take photos of the tea-pickers coming back from the fields, which took longer than expected but gave us some wonderful photos. Some of the women were happy to pose for a picture, while others refused point blank to have their photo taken – something we would experience repeatedly over the next few days.

We saw the first Indian rhinos in the distance as soon as we arrived at the park, which made us all the more hopeful that we would be able to get even closer at some point. The lodge in Kaziranga was set in beautiful surroundings and the accommodation was quite basic and traditional. Unfortunately, there was no cold beer or Coke to drink, so that evening we ended up sharing three chilled and three warm bottles of soda water between us ...

We were up and out early again the next day for our elephant safari, leaving for the park before breakfast in two off-road vehicles. We were seated in pairs behind a mahout on the elephants back and had to hold on tight as elephants sway quite vigorously when they walk. And the back of an elephant, even an Indian elephant, is decidedly wide, which means they are not all that comfortable ... but we were there to take pictures, of course, not to get comfortable!

Junger Indischer Elefant.
Young Indian elephant.

Hinduracke.
Indian roller.

Sonnenuntergang.
Sunset.

Satz aufsprang und sofort wieder im hohen Gras verschwand – zu schnell für ein schönes Foto, schade. Wir konzentrierten uns wieder auf die Rhinos und sahen im Frühdunst auch noch eine Herde Zackenhirsche, bis es dann zurück zu den Jeeps und zur Lodge ging. Nachmittags folgte eine weitere Safari, bei der wir neben einigen Nashörnern auch Nashorn-Vögel zu sehen bekamen. Und, leider schon in der Dämmerung und nur kurz vor uns auf der Straße, einen weiteren Tiger. Für etwas Unmut sorgte die restriktive Handhabung der Gebühren: Pro Person und Elefanten-Safari und zweimal den Eintritt in den Park waren zusätzlich noch Entgelte für jede einzelne Kamera, in der Regel zwei pro Person, zu entrichten.

Der Besuch in einem anderen Sektor am nächsten Morgen begann spät, weil wir eine längere Anfahrt hatten und dann der Kauf der Permits dauerte und dauerte. Leider waren wir diesmal in einem Gebiet, wo der schon allgegenwärtige Dunst durch die hohe Luftfeuchtigkeit noch durch Rauch von kontrollierten Feuern verstärkt wurde. Aufnahmen mit Telebrennweiten wurden dadurch sehr schwierig. Auf der Rückfahrt „mussten" wir im Dorf stoppen: Am ersten Tag des Holi-Festes *(mehr zu diesem Fest im Kapitel*

Kaziranga

Sitting on the back of an elephant gave us superb views over the top of the elephant grass, which can grow up to five feet tall, and it wasn't long before we saw the first Indian rhinos, despite them being fairly well hidden. Up high isn't necessarily the most flattering perspective for these heavyweights, which look like something straight out of Jurassic Park. Our musings were promptly interrupted by a growl about 30 yards away. As we turned to look, a tiger leapt powerfully into the air before disappearing immediately back in the high grass – too fast for a good photo, unfortunately. We turned back to the rhinos, and also saw a herd of barasingha through the morning mists, before heading back to the vehicles and then the lodge. Another safari was scheduled for the afternoon, and we were fortunate enough to see some more rhinos and hornbills. We also saw another tiger on the road, but the sun was already setting and he was only in front of us for a few moments. The day was quite expensive due to the restrictive charges, which were the cause of some resentment. In addition to paying per person, per elephant safari, we also had to pay the park entrance fee twice as well as a charge for every single camera, and most of us were carrying two cameras each.

Our visit to another sector the next morning started somewhat later, both due to the long journey and the even longer delays in buying a permit. Unfortunately, this region is known for being particularly hazy due to the high humidity, which was made worse during our visit by the smoke from controlled fires. It made taking photos with a long-range zoom lens very difficult. As we headed back to the lodge, we simply had to stop in the village as it was the first day of Holi celebrations (see the chapter „The country and the people" to find out more). A pavilion had been set up on the school grounds and the whole place decorated with colourful garlands. The children were having so much fun throwing mud at one another, and the boys were just pushing each other straight into the mud without even bothering to throw it. This was interspersed with singing and dancing, and every now and then people would stop to pour clean

Pelikan im Abendlicht.

Pelican in the evening light.

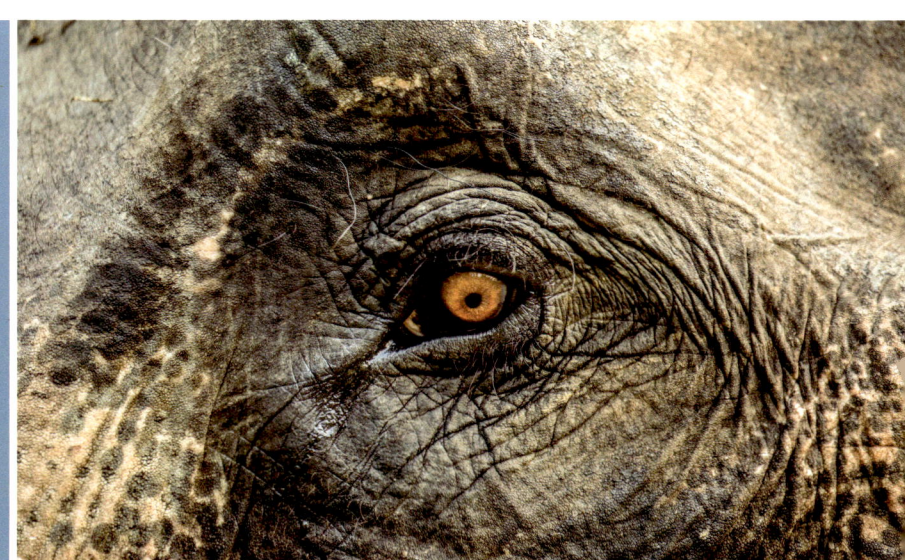

Auge eines Elefanten.

Close up of an elephants eye.

Kaziranga

"Land und Leute") war in einer Schule ein Festzelt aufgebaut und das Gelände mit bunten Girlanden geschmückt worden. Die Kinder, Jungen und Mädchen, hatten ihren Spaß daran, sich gegenseitig mit Schlamm zu bewerfen, die Jungs warfen andere auch direkt in den Matsch. Zwischendurch wurde getanzt und gesungen und auch mal klares Wasser über andere geschüttet. Unser fleißiges Fotografieren endete in einem beschleunigten Rückzug, als eine Gruppe der Mädchen sich zielstrebig in unsere Richtung bewegte!

Ein weiteres Erlebnis im östlichen Sektor Kazirangas ist eine Bootstour auf dem Brahmaputra, „Sohn des Brahma", und deshalb der einzige Fluss, der im Indischen den männlichen Artikel trägt. Ein Flusstal voller Sandbänke und bis zu 15 km breit! Eine große Gruppe Geier wollte vom Kadaver einer Kuh fressen, aber dieser lag fast ganz im Wasser, also war Warten angesagt. Vielleicht war das aber zu ihrem Vorteil, denn der reichliche Gebrauch von Diclofenac in der Tiermedizin führt in der Nahrungskette zur Aufnahme durch die Geier, die dann an Nierenversagen sterben – in solchen Mengen, dass diese früher in vielen Arten und zu Millionen vorhandenen Vögel nur noch in drei Arten vorhanden sind und als stark gefährdet eingestuft werden.

Unser anschließendes Mittagessen in einer Lodge wurde auch „farblich" gewürzt: Eine Gruppe von indischen Familien kam zu einem Picknick und feierte das Holi-Fest mit roten Wasserbomben. Allzu vorwitzige Fotografen wurden direkt mit einbezogen! Also wollten wir uns gegen wilde Attacken schützen, indem wir uns gegenseitig mit knalligen Farben bemalten – quasi eine Impfung mit aktiven Erregern, es stand nämlich ein Besuch eines nahegelegenen Dorfes an. Die Einheimischen hatten sich traditionelle Kleider angezogen und empfingen uns mit einer Weihrauchzeremonie, präsentierten Tänze und Musik sowie ihre Häuser und bestaunten ihrerseits diese „bunten Vögel" aus Europa. Ganz Mutige machten sogar versteckt ein paar Fotos mit ihren Handys von uns!

Ein Geier trocknet seine Flügel.

A vulture dries its wings.

Farbenprächtige Blüte vor dem Blattaustrieb.

Colourful blossoms before the first shoots.

water over each other. We ended up having to abandon our picture-taking and run back to the cars when some of the young girls noticed us and started heading in our direction!

A boat trip along the Brahmaputra River in the eastern sector of Kaziranga is also highly recommended. Brahmaputra means „son of Brahma", making it the only „male" river in India. The river valley is sculpted by sandbanks and is more than 9 miles wide in places. As we passed, a large group of vultures were hovering around the carcass of an almost fully submerged cow, waiting for the water level to drop. The widespread use of diclofenac by the veterinary profession is affecting vultures via the food chain. When they ingest the compound, it leads to kidney failure. These birds are dying in such large numbers that vulture populations – which were once in the millions – have now plummeted so dramatically that just three critically endangered species remain.

Afterwards, we visited another lodge for a very „colourful" lunch. A group of Indian families were there for a picnic, celebrating Holi with red water balloons and impertinent photographers were included without hesitation! Next on our agenda was a visit to a nearby village, so we decided to immunise ourselves against attack by painting each other with bright colours. The locals, wearing traditional dress, welcomed us with an incense ceremony, gave dance and musical performances, and showed us their homes, all the while marvelling at their brightly coloured visitors from Europe. A few were even brave enough to take some discreet photos of us with their mobiles!

Autor/Author: Harald Lydorf

Schwarze Drongos auf Futtersuche auf einem Nashornrücken.

Black drongos search for food on the back of a rhino.

Kleiner Jumbo unter Obhut.

An elephant calf in the care of an adult.

Täglich grüßt die Vogelvielfalt: Braunkopfliest (oben links), Silberklaffschnabel (oben rechts), fliegende Rostgänse (Mitte) und Nashornvogel (unten).

Diverse birdlife: White-throated kingfisher (top left); Asian openbill stork (top right); ruddy shelducks in flight (middle) and a hornbill (bottom).

Was für ein Größenunterschied: Sambarhirschkalb (oben) und Wasserbüffel (unten).

A sambar deer calf (top) is dwarfed by a water buffalo (bottom).

Zwei der größten Landsäugetiere überhaupt: Elefanten ...

Two of the world's biggest land mammals: Elephants...

... und Panzernashörner im Kaziranga Nationalpark.

... and Indian rhinos in Kaziranga National Park.

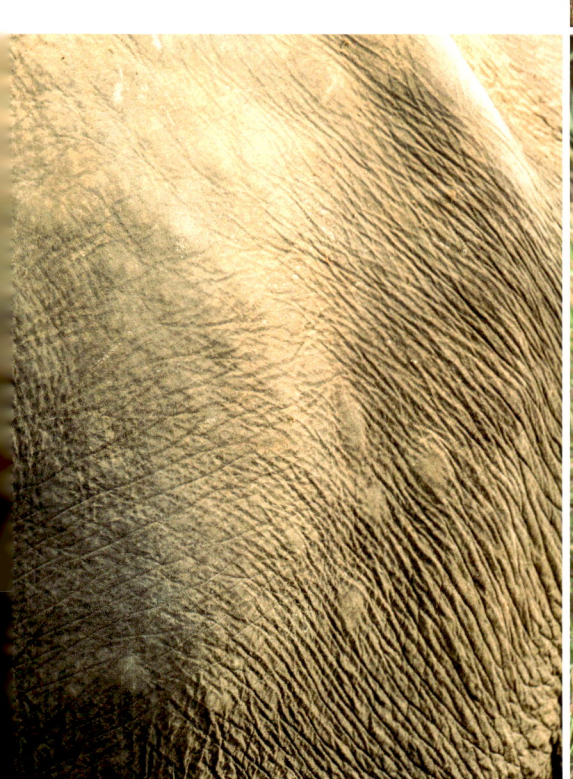

Gir

Heimat der letzten
Asiatischen Löwen

Gir

Home of the last
Asiatic lions

Asiatische Löwen – sie gehören zu den Letzten ihrer Art in Asien.

Asiatic lions are among the last of their kind in Asia.

Der Gir-Nationalpark liegt im indischen Bundesstaat Gujarat, der aufgrund seines restriktiven Alkoholverbotes auch der „trockene Staat" genannt wird. Dieses Naturreservat, auf der Halbinsel Kathiawar ganz im Westen Indiens gelegen, ist das Rückzugsgebiet der allerletzten wildlebenden Asiatischen Löwen. Die Wälder von Gir waren einst das Jagdgebiet des Nawabs von Junagadh. Der Asiatische Löwe, Sinnbild der Macht, wurde im 19. Jahrhundert zum beliebtesten Jagdwild indischer Herrscher und weißer Großwildjäger. Bereits 1880 war er mit Ausnahme des Gir-Forest in allen übrigen asiatischen Gebieten durch die Jagd, die Abholzung der Wälder und die Wilderei ausgerottet.

Schon 1900 wurde der Löwe durch den moslemischen Provinzfürsten von Junagadh unter Schutz gestellt, ist jedoch durch Wilderei weiter bejagt worden. Erst im Jahr 1965 wurde das Gir Wildlife Sanctuary als erstes Schutzgebiet in Gujarat eingerichtet. Mit der Umsiedlung eines Teils der einheimischen Viehhirten samt ihrer Herden und der Umwandlung der Kernzone des Reservates in einen Nationalpark begann 1975 der effektive Schutz der dort beheimateten Tiere. Heute umfasst der Nationalpark eine geschützte Kernzone und ein auch landwirtschaftlich nutzbares Schutzgebiet.

Dank seiner Nähe zum Arabischen Meer – die Küste ist nur ca. 40 km entfernt – durchzieht den Park eine ständige Brise. Auch wenn diese keine wirkliche Abkühlung bringt, so macht sie uns doch den Aufenthalt während der heißesten Jahreszeit erträglicher. An manchem Morgen sammeln sich an der Trennlinie von Land und Meer träge Wolken und schieben eine dunstige Schwüle von der Küste über die Landschaft. Das treibt uns die Schweißperlen auf die Oberlippe und lockt die Moskitos aus ihren Verstecken, sogar schon morgens um 5.30 Uhr, wenn die vollbesetzten Jeeps am Eingangstor zum Park auf Einlass warten.

Gir Nationalpark	
Bundesstaat	Gujarat
nächstgelegene Stadt	Junagadh
Größe [km²]	259
Schutzgebiet seit	1965
Nationalpark-status seit	1975
beste Reisezeit	Januar – Mai
Zoneneinteilung	1
Landschaft	Laubwald, Teakbaum, Dorngebüsch, Savanne
Tierwelt	
Raubtiere	Asiatischer Löwe, Leopard, Lippenbär, Goldschakal, Streifenhyäne
Grasfresser	Sambarhirsch, Axishirsch, Wildschwein, Nilgauantilope, Indische Gazelle, Vierhornantilope, Hirschziegenantilope
Vögel	Pfau, Eisvogel, Paradiesschnäpper, Wellenbrust-Fischuhu, Wollhalsstorch, Wespenbussard, Habichtsadler
Primaten	Hanuman-Langur
Andere	Sumpfkrokodil, Kobra

Gir

Gir National Park is located in the Indian state of Gujarat, often called the „dry state" due to its strict prohibition laws. This nature reserve stands on the Kathiawar peninsula in India's far west and is home to the sole remaining population of Asiatic lions living in the wild. Gir's forests were once the hunting grounds of the Nawab of Junagadh. A symbol of power, the Asiatic lion became the prized quarry both of Indian monarchs and white big-game hunters. Extensive hunting, deforestation and poaching meant that, by 1880, the lions were extinct in all regions of Asia except Gir forest.

In 1900, the Muslim ruler of the princely state of Junagadh declared the species protected. Despite this, poachers continued to hunt the lions and it was not until 1965 that Gir Wildlife Sanctuary was set up as the first protected area in Gujarat. A number of local cattle herders along with their herds were then resettled in the park in 1975 and the central part of the reserve turned into a national park, marking a turning point that brought effective protection for the indigenous animals. Today, Gir National Park comprises a fully protected core zone along with a protected area for agricultural use.

With the Arabian Sea just 25 or so miles away, the park enjoys a constant breeze that made our stay during the hottest time of year more bearable, if not exactly cool. Some mornings, clouds gather where the land meets the sea and the air turns hazy and humid as it moves inland. It was enough to draw beads of sweat to our

Strenge Einlasskontrollen huldigen der indischen Bürokratie und sollen den Stress für die Tiere regeln.

Indian red tape: Strict entry requirements are designed to keep the stress for the animals under control.

Gestärkt durch eine morgendliche Banane und mit reichlich Wasservorrat ausgestattet, fanden auch wir uns am noch verschlossenen Tor ein. Hier treibt die auf indische Art adaptierte europäische Bürokratie ihre seltsamen Blüten. Nicht nur dass jedem Jeep, identifizierbar anhand der Autonummer, ein fester Fahrer und Guide zugeteilt wird, auch die Gäste sind namentlich einem bestimmten Kennzeichen zugeordnet. Freie Platzwahl also Fehlanzeige!

Das hügelige Gebiet des Parks ist von vielen kleinen Tälern und Mulden durchzogen, in denen sich während des Monsuns das Regenwasser sammeln kann. Die lockeren Mischwälder und das Grasland bieten ca. 65.000 Huftieren wie Sambarhirschen, Axishirschen, Nilgauantilopen und Indischen Gazellen idealen Lebensraum. Daneben leben hier Wildschweine, Languren und Stachelschweine. Mit dem ansteigenden Bestand an Beutetieren war es möglich, dass sich außer den zur Zeit ungefähr 400 Asiatischen Löwen auch andere Raubtiere wie Leoparden, Streifenhyänen und Rohrkatzen im Gir-Nationalpark heimisch fühlen. Damit besitzt er eine der größten Raubtierdichten Indiens.

Ende Mai, kurz bevor der Monsun einsetzt, empfängt uns eine vollkommen ausgedörrte Landschaft. Bei Temperaturen von mehr als 40 Grad erwecken die

Kurz vor dem großen Regen prägen Ende Mai nahezu kahle Bäume und braune Erde den Gir-Nationalpark.

Shortly before the rains come, the landscape of Gir National Park is dominated by bare trees and brown earth.

brows and the mosquitoes out of hiding, even at 5:30 in the morning when the packed safari vehicles sat waiting at the park entrance for the gates to open.

After breakfasting on a banana and replenishing our water supplies, we also joined the queue in front of the locked entrance, where European bureaucracy at the hands of the Indians has yielded interesting results. Not only is each vehicle identified based on its registration number and allocated a specific driver and guide, but the passengers are also assigned by name to a specific registration number. Think you can sit where you like? Forget it!

The park's rolling landscape is contoured with many small valleys and depressions where rain water collects during the monsoon season. The loosely packed mixed forests and grasslands provide the ideal habitat for some 65,000 members of the ungulate family, including sambar deer, chitals (or spotted deer), nilgai (blue bull antelopes) and chinkara (Indian gazelles). They live alongside wild boar, langurs and Old World porcupines. Growing populations of prey has meant that other predators, such as leopards, striped hyenas and jungle cats, also feel at home next to the 400 or so Asiatic lions currently living in Gir National Park, giving it one of the densest populations of predators across all of India.

We arrived in late May, just before the monsoon season, and were welcomed by an extremely arid landscape. Temperatures soared to more than 40 degree centigrade, making the dry deciduous teak forests seem dead and lifeless. As we passed through, we occasionally saw a flowering royal Poinciana tree marking each of

Dieser Hirtenjunge gehört zur jungen Generation der Maldharis.

This young herdsman is part of the Maldhari community's next generation.

trockenen laubabwerfenden Teakwälder einen leblosen abgestorbenen Eindruck. Hier und da markiert ein blühender Flamboyantbaum wie eine Fackel die angrenzenden Dorfgemeinschaften. Der Wind in den letzten Blättern erzeugt ein Geräusch wie von leise raschelndem Papier. Mitten in diesem Staubgrau ziehen sich vereinzelt schmale grüne Streifen durch die Pflanzenwelt. Sie künden entweder von den letzten Rinnsalen der sonst vorhandenen Flüsse oder sie verlaufen am Ufer des einzigen ganzjährig wasserführenden Flusses Hiran. Für uns ist es kaum vorstellbar, dass in ca. drei Wochen, wenn der Monsun einsetzt, hier alles wieder grünt und blüht. Den kalkweißen „Geisterbäumen", deren Baumrinde nachts im Mondlicht schimmern soll, wird der Regen aber auch kein Leben mehr einhauchen können.

Auf eingangs fest zugewiesenen Routen begeben sich die Guides mit uns in einer ebenfalls fest vorgeschriebenen Zeit auf „Löwenjagd". Dabei hängt es nicht immer von Fortuna ab, ob die Raubtiere gesichtet werden. Obwohl von der Parkleitung untersagt, wird das eine oder andere Mal mit traditionellen Treibermethoden dafür gesorgt, dass die Tiere ihr schützendes, wenig fotogenes Versteck im Unterholz verlassen. Aus der Sicht der Guides, die ihr knappes Monatsgehalt (ca. 9.000,00 Rps. = 110,00 €) aufstocken wollen, verständlich – für die Tiere jedoch bedeutet

Zu den mehr als 250 Vogelarten im Park gehört auch dieser kleine Kerl.

This little chap is one of the more than 250 species of bird found in the park.

Gir National Park	
State	Gujarat
Nearest city	Junagadh
Size [sq mi]	100
Protected since	1965
National Park since	1975
Best time to travel	January – May
Zones	1
Landscape	Deciduous forest, teak trees, thorn bushes, grassland
Fauna	
Predators	Asiatic lion, leopard, sloth bear, golden jackal, striped hyena
Grazing animals	Sambar deer, chital (spotted deer), wild boar, nilgai (blue bull antelope), chinkara (Indian gazelle), four-horned antelope, blackbuck
Birds	Peafowl, common kingfisher, paradise flycatcher, brown fish owl, woolly-necked stork, European honey buzzard, the Bonelli's eagle
Primates	Grey (hanuman) langur
Other	Mugger crocodile, cobra

Ein seltenes Glück ist die Begegnung mit einem scheuen Leoparden.

A rare encounter with the reclusive leopard.

the neighbouring village communities like a flaming torch. The wind blowing through the last remaining leaves made a sound like gently rustling paper. This dusty grey landscape was crossed in places by narrow green ribbons slicing through the earth. These were either the last remaining rivulets of what, in wetter times, are full-sized rivers, or they marked out the banks of the only year-round water-bearing river, the Hiran. It was hard for us to imagine that the monsoon would transform this view into a green and thriving landscape in just three weeks or so. But no amount of rain would be enough to bring life back to the chalky white „ghost trees", whose bark is said to shimmer by moonlight.

The guides set out along their rigidly assigned routes to take us on our „lion hunt" with closely regulated hours. Seeing the predators in their natural habitat is not necessarily always down to fate, however. Although strictly forbidden by the park authorities, the guides sometimes use traditional herder methods to drive the animals out of their safe hiding places in the undergrowth, away from the cameras. The guides are of course keen to top up their scant wages (about 9,000 rupees, or 110 euros per month), but it means more stress for the animals. Because we are on a tight schedule, however, we only stopped for a few minutes each time before heading on to find the park's next residents.

The national park is also home to some 250 bird species, making Gir forest a veritable Garden of Eden for ornithologists and birdwatchers. Perhaps the most striking and colourful bird they come to see is the national bird of India, the Indian peafowl. The male is known for its elongated train of upper-tail feathers that can reach up to five feet in length and were once used to adorn the King's throne. Astonishingly, their iridescent plumage is completely devoid of pigmentation. The colours are created by intricate, microscopic air

es zusätzlichen Stress. Aufgrund des Zeitreglements bleiben trotzdem meist nur wenige Minuten, bevor es auf die Suche nach weiteren Parkbewohnern geht.

Mit etwa 250 Vogelarten, die der Nationalpark beherbergt, ist der Gir-Forest ein echtes Eldorado für Ornithologen und Vogelbeobachter. Der Nationalvogel der Inder, der Asiatische Blaue Pfau, ist dabei wohl der auffallendste und farbenprächtigste Vogel. Die Schleppe der Männchen besteht aus ein bis eineinhalb Meter langen Oberschwanzfedern, die einst Königsthrone schmückten. Das schillernde Gefieder enthält erstaunlicherweise keine Farbpigmente, der Farbeindruck entsteht durch Interferenz des Lichts in mikroskopisch kleinen Luftkammern der Federn. Zerstört man die Luftkammern, sind die Federn dunkelgrau. Dieses fasanenartige Tier ist aufgrund seines stark entwickelten Geruchs- und Gehörsinns die beste Alarmanlage bei Gefahr. Seine lauten durchdringenden Schreie liegen wie ein Geräuschteppich über der Gegend und begleiten uns an jedem Tag im Park.

In enger Gemeinschaft zur Natur und zu den Wildtieren lebt auch heute noch ein großer Teil der Volksgruppe der Maldharis im Park. Dieses halbnomadische Hirtenvolk hütet Kühe, Schafe und Büffel sowohl im als auch rund um den Park. Häufig „bezahlen" sie dafür mit Viehverlusten an die Raubkatzen. Ihren Lebensunterhalt bestreiten sie als traditionelle Milchbauern mit Milch und Käse.

Es erstaunt uns, wie „bewohnt" der Nationalpark ist. Denn außer den Maldharis lebt hier ein großer Teil der ethnischen Gruppe der Siddi. Das sind Einwanderer mit afrikanischen Wurzeln, die seit dem Mittelalter vorwiegend als Sklaven nach Indien kamen. Einige ihrer Vorfahren dürften aber auch als freie Soldaten, Seefahrer oder Händler gekommen sein. Neben ihrem Kraushaar und dem dunkleren Hauttyp sind deutliche Zeichen ihrer afrikanischen Herkunft in Tanz und Musik erhalten geblieben.

Darüber hinaus findet sich im Nationalpark noch der Schrein des Kankai Mata, der als Wohltäter der Hirten des Gir-Waldes gilt. Die Tempelmauern gelten als sicher und täglich kommen Gläubige und Touristen mitten in das Herz des Dschungels.

Den Sambar- oder Pferdehirsch machen seine großen Ohren zu einem echten Hingucker.

The large ears of the sambar deer help it to stand out.

Die schrillen Schreie des Asiatischen Blauen Pfaus sind regelmäßig und weit zu hören.

The piercing cries of the Indian peafowl can be heard often and from far away.

pockets in the feathers that filter and reflect light. If these air pockets are destroyed, the feathers are a dark grey. With its keen hearing and strongly developed sense of smell, it's hard to imagine a better „watchdog" than this member of the pheasant family. Its loud, penetrating cries are a permanent background accompaniment to our daily park visits.

Even today, a large community of Maldhari tribesmen still lives in harmony with nature and the wild animals whose park habitat they share. These semi-nomadic herders breed cattle, sheep and buffalo both in and around the park, often losing livestock to the wild cats as the price for living where they do. Dairy farmers by tradition, they earn their livelihood by producing milk and cheese.

We were surprised by how „lived in" the national park is. The Maldhari are not the only indigenous peoples to live there. A large Siddi community calls the park home, too. The Siddi are descended from the Africans that were brought to India mostly as slaves from as far back as the Middle Ages, while others probably came under their own volition as soldiers, sailors and merchants. In addition to their curly hair and darker skin, strong elements of their African heritage have been preserved in the form of dance and music. A shrine to Kankai Mata, benefactor of the shepherds of Gir forest, can also be found in the national park. The temple walls ensure visitors' safety and draw the faithful and tourists alike to the heart of the jungle on a daily basis.

Autor/Author: Kerstin von Splényi

Tarnung ist alles! Fast nicht zu erkennen sind die beiden Mungos.

Camouflage is everything! These two mongooses are hardly distinguishable from their surroundings.

Hanuman-Languren zählen in Indien als heilige Tiere zu den bekanntesten Affenarten.

Grey langurs are considered holy in India and are one of the most well-known simians.

Die Fahrer finden auch kleinste Eulen am Wegesrand (rechts), vereinzelt stehende Teakbäume erscheinen wie Wegemarken (unten).

Our drivers spot even the tiniest owl by the tracks (right); solitary teak trees appear like way posts (bottom).

Der Asiatische Löwe, Sinnbild der Macht, kämpft heute ums Überleben. Nur noch rund 400 Tiere leben heute im Gir-Nationalpark.

The Asiatic lion, symbol of power, is now fighting for its very survival. Just 400 or so of these animals currently live in Gir National Park.

Kanha

Wogendes Grasland und „europäische" Wälder

Kanha

Undulating grasslands and "European" forests

Kanha

Der Kanha-Nationalpark liegt im zentralindischen Hochland im Osten des Bundesstaates Madhya Pradesh mitten im Herzen Indiens und ist einer der bekanntesten Parks des Landes. Dieses großartige Tierreservat erstreckt sich über rund 940 km² mit Laubwäldern, Savannengrasland, Hügeln und sanft mäandrierenden Flüssen. Die Forstgebiete setzen sich neben Sal- und Teakbäumen auch aus feuchten Laubwäldern zusammen, die uns verblüffend stark an die Wälder Nordeuropas erinnern.

Welch ein Unterschied zum Gir-Nationalpark. Auch in Kanha ist es Ende Mai trocken und staubig, aber die Bäume sind grün und der Bambus steht meterhoch an den Wegen. Bambus gehört zu den schnell wachsenden Gehölzen. Beeindruckend ist, dass die meisten Bambusarten nur einmal in ihrem Vegetationsleben blühen und Früchte bilden, danach sterben sie ab. Damit nicht genug; sie blühen synchron periodisch in einer Region wie zum Beispiel ganz Europa und das je nach Art nur alle 12 bis 120 Jahre. Warum das so geschieht, ist bisher noch nicht eingehend erforscht. Die Natur birgt eben doch noch ihre kleinen Geheimnisse.

Für die „Jagd" nach dem besten Tiermotiv bietet die Periode der Wasserknappheit die größte Chance auf Erfolg. Denn ganz egal ob Pflanzen- oder Fleischfresser, ob Jäger oder Gejagter, jedes Tier muss mindestens einmal am Tag zu einem der wenigen noch verbliebenen Wasserlöcher

Kanha Nationalpark	
Bundesstaat	Madhya Pradesh
nächstgelegene Stadt	Jabalpur
Größe [km²]	940
Schutzgebiet seit	1933
Nationalpark-status seit	1955
beste Reisezeit	November – Juni
Zoneneinteilung	4
Landschaft	Grasflächen, Flusstäler, Salbaum, Teakbaum, Laubmischwald
Tierwelt	
Raubtiere	Bengaltiger, Leopard, Lippenbär, Mungo, Rothund
Grasfresser	Nilgauantilope, Sambarhirsch, Axishirsch, Gaur (Wildbüffel), Hochland-Barasinghahirsch, Hirschziegenantilope, Stachelschwein, Zwerghirsch, Muntjak (Bellhirsch)
Vögel	Nashornvogel, Bienenfresser, Indische Blauracke, Pirol, Hainparadiesschnäpper, Silberreiher, Eisvogel
Primaten	Hanuman-Langur, Rhesusaffe
Andere	Python

Die großen, schlank gebauten Hanuman-Languren haben ausgeprägte Überaugenwülste und ein haarloses Gesicht – ihr Schwanz kann bis zu 110 Zentimeter lang werden.

Grey langurs are tall and lean with heavy brow ridges and hairless faces. Their tails can reach up to 43 inches in length.

Kanha

Kanha National Park is one of the country's most famous parks. It lies in the central Indian highlands in the eastern part of the state of Madhya Pradesh in the heart of India. This amazing wildlife preserve boasts some 360 or so square miles of deciduous forests, grasslands, hills and gently meandering rivers. The forested regions are made up of sal and teak trees, as well as moist deciduous forests that reminded us very much of the forests of northern Europe.

What a contrast to Gir National Park! Our visit in late May meant it was dry and dusty in Kanha, too, but the trees were green and the tracks lined with bamboo standing tens of feet tall. Bamboo is an impressive and curious plant that grows extraordinarily fast. Most types of bamboo flower and bear fruit just once in their lifetime before dying. Even more astonishing is that they flower synchronously at specific intervals across entire regions – all of Europe, for example – and, depending on type, these intervals can be anything from 12 to 120 years. As yet, no definitive explanation has been found for why this happens. Mother Nature is keeping this secret for herself.

The best time to „hunt" for animal photos is when water is scarce. Regardless of whether herbivore or carnivore, hunter or hunted – every animal has to come out of hiding at least once a day to drink from one of the few remaining water holes. The only question is: which one? It's

Die Gaure, größte lebende Vertreter der Rinder, lassen sich gern von kleinen Helfern bei der täglichen Hygiene unterstützen.

The largest living bovine, gaur, are happy for their little helpers to assist with grooming.

Kanha

kommen. Die große Frage ist nur: zu welchem!? Die beiden Flüsse des Parks, Halon und Banjar, bilden hier und da größere und kleinere Wasserpools oder Bachläufe, die immer einen Fotostopp wert sind.

In den von uns besuchten Nationalparks muss bei jeder Safari, die im Allgemeinen bei Öffnung der Parks gegen 5.30 Uhr startet, das nicht immer klar durchschaubare Einlass-Prozedere mit viel Geduld und indischer Gelassenheit absolviert werden. Ausländische Touristen müssen sich bei jeder Einfahrt (also zweimal am Tag) mit ihren Passdaten registrieren lassen, daher ist der Reisepass der Gegenstand in unserem Reisegepäck mit den meisten Gebrauchsspuren. Kanha ist in vier Zonen eingeteilt und durch alle führt ein Netz befahrbarer Pisten. Welche Tiere aus den offenen Jeeps zu sehen sind, hängt dadurch in erster Linie von der Gebietswahl des Guides ab. Unsere Begleiter sind immer informiert, was wo am Vortrag gesichtet wurde ...

Eine Besonderheit Kanhas ist das letzte wildlebende Vorkommen des Hochland-Barasinghas, einer Hirschunterart, die nur hier beheimatet ist und in den 60er Jahren nur knapp vor der Ausrottung bewahrt werden konnte. Der Hochland-Barasingha lebt, abweichend von den anderen Barasingha-Unterarten, in Waldgebieten mit festem Untergrund. Die Bezeichnung Barasingha, ein Wort aus dem Hindi, das „zwölf Hörner" bedeutet, verweist auf das stark verzweigte Geweih, das häufig zwölf und mehr Enden aufweist.

In dieser urtypischen, an die Erzählungen Rudyard Kiplings erinnernden Landschaft, breiten tausend Jahre alte Bäume ihre Kronen weit aus und spenden nicht nur Schatten, sondern sind vielen exotischen Vogelarten wie indischen Blauracken, Bienenfressern, Eisvögeln oder seltenen Nashornvögeln Unterschlupf. Scharen von schwarzgesichtigen Languren bevölkern die Wälder. Gaur, Indiens größte Wasserbüffel, und Lippenbären gehören ebenso wie Leoparden und Rothunde zu den großen Säugetieren des Parks. Vielleicht hat ja eine Begegnung mit einem indischen Mungo Rudyard Kipling zu seinem Rikki-Tikki-Tavi aus dem Dschungelbuch inspiriert. Diese tagaktiven Tiere, die

Eine Tigersichtung wird oftmals zum Großereignis. Den Tiger scheint es nicht zu stören.

A tiger sighting is often a major event. The tiger seems unruffled.

Kanha National Park	
State	Madhya Pradesh
Nearest city	Jabalpur
Size [sq mi]	363
Protected since	1933
National Park since	1955
Best time to travel	November – June
Zones	4
Landscape	Grassy plains, river valleys, sal trees, teak trees, mixed deciduous forest
Fauna	
Predators	Bengal tiger, leopard, sloth bear, mongoose, dhole (Indian wild dog)
Grazing animals	Nilgai (blue bull antelope), sambar deer, chital (spotted deer), gaur (Indian bison), barasingha (hard ground swamp deer), blackbuck, Old World porcupine, dwarf musk deer, muntjac (barking deer)
Birds	Hornbill, bee eater, Indian roller, Euroasian golden oriole, Asian paradise flycatcher, great egret, common kingfisher
Primates	Grey (hanuman) langur, rhesus macaque
Other	Python

always worth stopping for a photo opportunity at one of the many natural depressions where water from the Halon and Banjar rivers that run through the park collects in pools of different sizes or where the rivers break off to form streams.

In the national parks we visited, the safaris generally started when the parks opened at 5:30 am. The complex procedure for getting into the park was mostly a waiting game that called for some typical Indian aplomb. Foreign tourists have to register with their passport details every time they enter (twice a day), so our passports were fairly battered and dog-eared by the end of it. Kanha is split into four zones and criss-crossed with tracks accessible to vehicles. The animals we spotted from the open safari vehicles depended mainly on where the guides took us, and they always knew which animals had been seen where the day before ...

Kanha is particularly unique as it is home to the last population of barasingha, or hard ground swamp deer, living in the wild – a species that was close to extinction in the 1960s. Unlike other members of this deer family, the barasingha has adapted to living on hard ground in forested areas. The name barasingha, which means „twelve-tined" in Hindi, refers to its heavily tined antlers, which can carry 12 or more.

Der wichtigste Posten ist der auf dem Ausguck, davon kann das Leben abhängen.

The look-out is the most important job, as it can often mean the difference between life and death.

Die sanften Riesen scheinen vollkommen mit ihrer Umgebung zu verschmelzen.

These gentle giants seem to blend entirely into the landscape.

auch hier immer wieder mal kurz vorbeihuschen, treten in indischen Fabeln als Beschützer der Menschheit auf, in dem sie vor Schlangenangriffen schützen.

Hauptattraktion in Kanha sind unumstritten die Tiger. Ganz Indien nimmt am Wohl und Wehe der Tiger teil. Anders als in Afrika strömen neben ausländischen auch einheimische Touristen zu Tausenden in die Nationalparks. Die meisten Bengaltiger in Indien haben sogar einen Namen oder zumindest eine Nummer, die sie eindeutig identifizieren. Frei nach Saint-Exupéry fühlt sich jeder nur für das verantwortlich, was er sich vertraut gemacht hat.

Die Nähe der Raubkatze ist förmlich spürbar. Die angespannte Stille im Wald überträgt sich auf Mensch und Tier. Mit größter Aufmerksamkeit verharren die Besucher, die Beutetiere und sogar die Vögel. Vereinzelte Warnrufe zeigen an, wo sich die großen Jäger aufhalten. Bis plötzlich ein dumpfes Grollen im langsam schon schwächer werdenden Sonnenlicht allen Anwesenden eine Gänsehaut in den Nacken treibt. Das Gebrüll eines Tigers kündet von Kraft, Urinstinkten und absolutem Herrschaftsanspruch. Ein Klang, den man nie wieder vergisst.

Bis 1933 war das Kanha-Tal das Jagdrevier des Vizekönigs, das hochrangigen britischen Offizieren und Staatsbeamten vorbehalten war. Nationalpark wurde Kanha allerdings erst 1955 und gehörte zu den ersten Teilnehmern am Projekt Tiger, das von der indischen Regierung

Kanha's distinctive landscape calls to mind the stories of Rudyard Kipling. Thousand-year-old trees spread their canopies wide and offer not only shade but sanctuary to many species of exotic bird, such as the Indian roller, bee eaters, the common kingfisher, or the rarer hornbill. Troops of black-faced grey langurs patrol the forests. Gaur, the largest bovine found in India, and sloth bears make up the large mammals in the park, along with leopards and dholes, or Indian wild dogs. Perhaps an encounter with an Indian mongoose inspired Rudyard Kipling to come up with The Jungle Book's Rikki-Tikki-Tavi. These diurnal animals scampered past us regularly and often crop up in Indian fables as guardians of humans that protect against attacks by snakes.

Of course, Kanha's biggest attraction are the tigers. The well-being of the tiger population is close to the heart of everyone in India. Unlike in Africa, Indians themselves also flock to the national parks in their thousands, alongside the foreign tourists. Most Bengal tigers in India even have a name or at least a number, to identify them. To paraphrase Antoine de Saint-Exupéry, you become responsible for whatever you have named.

The proximity of these big cats was palpable. The tense silence over the forest gripped humans and animals alike. Visitors, prey – even the birds paused attentively. Sporadic warning calls revealed the whereabouts of the big hunters, before, out of the blue, a low rumbling in the slowly fading sunlight caused the hairs on the backs of our necks to stand up. The roar of a tiger is a sound of power, primal instinct, and supreme dominance. It is a sound you never forget.

Until 1933, the Kanha valley was the hunting ground of the Viceroy of India and reserved exclusively for use by high-ranking British officers and state officials. It wasn't until 1955 that Kanha gained national park status, before becoming one of the first parks to participate in the Project Tiger initiative launched by the Indian government under Indira Gandhi. The project is run by the Ministry of Environment and Forests and focuses on establishing tiger reserves using a strategy of core and buffer zones. The core areas are awarded national park status, which means the animals living there are fully protected. The peripheral zones are developed as multiple- and mixed-use areas to create a buffer to human settlement areas and to protect the interests of the indigenous population.

In einem geschützten Versteck widmet sich der Tiger seinem Riss. Tiger beginnen meist am Hinterteil zu fressen, während Löwen in der Regel zuerst die Bauchhöhle öffnen.

The tiger retreats to its hideout to devour its prey. Tigers usually eat the hindquarters first, while lions often start by tearing open the stomach cavity.

Kanha

Der Axishirsch trägt sein Fleckenkleid ein Leben lang. Bauch und Beine sind weiß gefärbt.

The chital wears its spots a lifetime long. Its underbelly and legs are white.

Im Kanha-Nationalpark wird erfolgreiches Management mit der Erforschung der Tierwelt verbunden. Dafür wurden beispielsweise mehr als 50 Camps für Ranger eingerichtet, die als Fußpatrouillen täglich unterwegs sind. Mehr als einmal begegneten uns die spärlich bewaffneten Männer auf ihren alten Fahrrädern. Ungefähr 8 bis 10 Kilometer legen sie täglich zurück. Teilweise ermöglicht ihnen die moderne Technik eine drahtlose Kommunikation untereinander, die im Ernstfall lebensrettend sein kann. Während des Monsuns, wenn die Wege mit Jeeps und zu Fuß unpassierbar sind, kommen Elefanten-Patrouillen zum Einsatz. Es scheint eine schier unlösbare Aufgabe, die Interessen der sich stetig ausdehnenden Bevölkerung und der zu schützenden Wildtiere in Einklang zu bringen. Doch die Bemühungen der letzten Jahrzehnte, die auch auf Brand- und Gewässerschutz erweitert wurden, zeigen erste Erfolge, sowohl im Tierbestand als auch im Verhalten der umliegenden Bewohner. Seither hat sich die Zahl der Tiger in Kanha wieder auf über 200 Tiere erhöht.

unter Indira Gandhi initiiert wurde. Dieses Projekt des Ministeriums für Umwelt und Forstwirtschaft baut Tigerreservate auf einer Kern- und Pufferzonenstrategie auf. Die Kernzonen erhalten den Status eines Nationalparks, der es erlaubt, die darin lebenden Tiere unter absoluten Schutz zu stellen. Die Pufferzonen hingegen werden als Bereiche mit Mehrfach- und Mischnutzung entwickelt, um so einen Übergang zum menschlichen Siedlungsgebiet zu schaffen und die Interessen der einheimischen Bevölkerung zu wahren.

Der Schakal ist deutlich kleiner als ein Wolf und ein ausdauernder Läufer.

The jackal is much smaller than the wolf and is a long-distance runner.

Kanha National Park successfully combines management with animal research projects. For example, it has set up more than 50 camps for rangers who go out on patrol every day. Several times we encountered meagrely armed men on old bicycles who travel around 5 to 6 miles every day. Occasionally, they are able to communicate with one another wirelessly thanks to modern technology, which could be life-saving for them in an emergency. During monsoon season, when the tracks are impassable for vehicles or on foot, the elephant patrols come out. Balancing the interests of the ever growing human population with those of the wild animals that need protecting seems like an impossible task. Efforts in recent decades, however, which have been expanded to include fire protection and water conservation activities, are starting to bear fruit. Animal populations are beginning to recover and local residents are becoming more aware. The number of tigers in Kanha has now risen to more than 200.

Autor/Author: Kerstin von Splényi

Tausend Jahre alte Bäume breiten ihre Kronen aus wie ein Dom.

Thousand-year-old trees spread their cathedral-like canopies.

Der Barasingha oder Zackenhirsch ist ein nur in Indien lebender Hirsch, der wie hier auch Symbiosen eingeht (linke Seite).

The barasingha, or swamp deer, is a species that only lives in India and forms symbioses with other living creatures (far left).

Wenn der Tiger auf der Jagd durch sein Revier streift, fungiert der Barasingha mit seinem Schrecklaut, einem hohen Bellen, als Gefahrenanzeiger.

The barasingha's loud alarm call acts as a warning when the tiger is out hunting for prey.

Die farbenprächtige Hinduracke (rechts).

The colourful Indian Roller (right).

Was die beiden Langurenmännchen wohl zu bereden haben (unten)?

What are these two langurs chatting about (bottom)?

Mystische Morgenstimmung und einsame Waldwege im Kanha-Nationalpark.

A magical morning and secluded forest tracks in Kanha National Park.

Äußerst selten trifft man zwei Tiger zusammen an. Da ist vorsichtiges Beschnuppern angesagt ...

Two tigers together is an extremely rare sight. They start with some cautious sniffing ...

Laut Aussage unseres Guides sitzt hier eine „Dschungelkatze" im Gras.

Our guides tell us that a "jungle cat" is sitting in the grass.

Durchaus kein Kuscheltier ist der Lippenbär mit seinem zotteligen schwarzen Fell und den unbehaarten Lippen.

The sloth bear is no cuddly toy, with its shaggy black coat and hairless lips.

Gaure gehören zu den fünf Rinderarten, die von Menschen domestiziert wurden.

Gaur are one of five kinds of bovine to have been domesticated by humans.

Bandhavgarh

Im Reich des Bengalischen Tigers

Bandhavgarh

Home of the Bengal tiger

Bandhavgarh

Inmitten des Vindhya-Gebirges im zentralindischen Staat Madhya Pradesh liegt der Bandhavgarh-Nationalpark, dessen Name vom höchsten Berg (807 m) dieser Gegend stammt. Die hügelige Landschaft wird von einem Plateau überragt, auf dem einst das Fort der Maharajas stand. Der Überlieferung nach führt der Bau des Forts zurück in die Zeit des Ramayana, neben dem Mahabharata das zweite indische Nationalepos. Es heißt, der Affenkönig Sugriva und sein Minister Hanuman schufen diesen Ort für den Prinzen Rama, um sich nach seinen Schlachten auszuruhen. Für Besucher ist das Fort derzeit geschlossen.

Bandhavgarh Nationalpark	
Bundesstaat	Madhya Pradesh
nächstgelegene Stadt	Umaria
Größe [km²]	449
Schutzgebiet seit	
Nationalparkstatus seit	1968
beste Reisezeit	Mitte Okt. – Anfang Juni
Zoneneinteilung	5
Landschaft	Salbaum, Mischwald, Grasland
Tierwelt	
Raubtiere	Bengaltiger, Leopard, Lippenbär, Goldschakal, Streifenhyäne, Mungo, Rohrkatze, Wildhund
Grasfresser	Muntjak (Bellhirsch), Sambarhirsch, Axishirsch, Gaur (Wildbüffel), Wildschwein, Stachelschwein, Vierhornantilope, Indisches Schuppentier
Vögel	Bankivahuhn, Weißnackenspecht, Langschnabelgeier, Storch, Fischuhu, Nashornvogel, Adler, Falke, Fliegenschnäpper
Primaten	Rhesusaffe

Bandhavgarh war der Sitz mehrerer indischer Dynastien und herrschaftliches Jagdgebiet. Im näheren Umkreis befinden sich wildreiche Grasgebiete, die aus Sümpfen hervorgegangen sind, die ihrerseits einst zum Schutz des Forts angelegt wurden. Nach der Unabhängigkeit Indiens blieb Bandhavghar zunächst in Privatbesitz, bis das Gebiet im Jahr 1968 durch den Maharaja von Rewa an den indischen Staat als Schutzgebiet übereignet wurde. Heute ist der Park in fünf Zonen aufgeteilt (Tala, Magdhi, Kallwah, Khitauli und Panpatha) und ist eines der am besten gemanagten Schutzgebiete Indiens, wobei die örtliche Bevölkerung aktiv mit einbezogen wird. Die tiefen Täler sind mit Teakholz- und Bambuswäldern bedeckt. Das offene Grasland und die Mischwälder bieten ausgezeichnete Möglichkeiten für Wild- und Vogelbeobachtungen.

Die meisten Jeepsafaris führen in die Tala-Zone, wo die Chance auf eine Tigersichtung aufgrund der übersichtlichen Größe und einer beachtlichen Tigerpopulation am größten ist. Sogar weiße Tiger soll es hier geben, das letzte Mal wurde 1951 ein Exemplar von Maharaja Martand Singh gefangen. Die meisten unserer

Manche Gehölze bilden seltsame Formen.

The trees can grow into some strange shapes.

Bandhavgarh

Set among the Vindhya mountains in the central Indian state of Madhya Pradesh is Bandhavgarh National Park, which derives its name from the highest hill in the area (2,650 ft). The hilly landscape is dominated by a plateau where the former fort of the maharajas stands. Construction of the fort is said to go back to the time of the Ramayana, one of the two great Indian epics, the other being the Mahabharata. Legend has it that Sugriva, king of the monkeys, and his minister Hanuman constructed it for prince Rama as a place where he could recuperate from his battles. The fort is currently closed to tourists.

Bandhavgarh was the seat of several Indian dynasties and a hunting ground of the ruling classes. Nearby are game-rich grasslands, which developed from the swamps that were once created to protect the fort. After India gained its independence, Bandhavghar initially remained in private ownership until the Maharaja of Rewa handed the area over to the state as a nature reserve in 1968. The park is currently divided into five zones (Tala, Magdhi, Kallwah, Khitauli and Panpatha) and is one of the best managed nature reserves in India, with the local population actively involved in its running. The deep valleys are home to teak and bamboo forests, while the open grassland and mixed forests provide excellent opportunities for observing birds and game.

Most safaris take tourists to the Tala zone, where the chances of seeing a tiger are greatest due to the modest size of the zone and the significant tiger population. There are supposedly even white tigers here and the last time one was caught was by Maharaja Martand Singh in 1951. Most of our photos of tigers were taken here and in Ranthambore National Park. Full-body shots, close-up portraits, prostrate, standing, walking, drinking – it's almost like we were legendary big game hunters; we may not have sat round the campfire in front of our tents, but we proudly presented our photographic trophies after dinner.

It could have far-reaching consequences here if guides and guests were to lose track of time in all the excitement, becau-

„Die Geier warten schon" – ihre Flügelspannweite kann mehr als 2 Meter betragen.

"The vultures are circling" – their wings have been known to span more than six-and-a-half feet.

Bandhavgarh

Tigeraufnahmen entstanden hier und im Ranthambore-Nationalpark. Ganzkörperaufnahmen, Close-up-Porträts, liegend, stehend, gehend, trinkend – fast ist es wie bei den legendären Großwildjägern. Wenn auch nicht am Lagerfeuer vor dem Zelt, so doch nach dem abendlichen Dinner werden die fotografischen Trophäen voller Stolz präsentiert.

Sollten Guides und Gäste im Eifer des Gefechtes die Zeit aus dem Auge verlieren, kann das fatale Folgen haben. Denn die einzuhaltenden Ausfahrtszeiten sind hier, wie in allen von uns besuchten Parks, streng festgelegt. Bei Nichteinhalten dieser Zeiten kann den Fahrern und Guides eine Arbeitssperre von bis zu sechs Monaten drohen. Zum Glück ist Indien aber auch das Land der Geschichtenerzähler und so fällt den cleveren Guides am Ende einer staubigen und halsbrecherischen Ausfahrtsrallye immer noch die eine oder andere sagenhafte Geschichte als Entschuldigung ein.

Begegnungen mit Rotwild wie Muntjaks, Nilgau-Antilopen oder Axishirschen sowie mit Rhesus- oder Languren-Affen sind nahezu garantiert. Aber auch Lippenbären, Stachelschweine und Sambarhirsche leben im Wald verborgen, während sich Hyänen, Füchse und Schakale gelegentlich auch im offenen Gelände zeigen. Daneben kommen etwa 200 Vogelarten im Park vor. Mittlerweile wurde auch der Gaur nach seinem unerklärlichen Verschwinden Ende der 90-er Jahre des vorigen Jahrhunderts erfolgreich wieder angesiedelt. Der Gaur ist der größte lebende Vertreter der Rinder. Ein Gaurbulle erreicht die beeindruckende Kopf-Rumpf-Länge von 3,30 m, eine Körperhöhe von 2,20 m sowie ein Gewicht von über einer Tonne. Diese als gefährdet eingestuften Tiere leben in hügeligem Gelände und steigen gebietsweise bis 1.800 m auf. Es sind typische pflanzenfressende Gemischtköstler, die sowohl Gras als auch Laub und Kräuter zu sich nehmen.

Unter den Säugetieren des Parks steht neben dem Tiger der Lippenbär auf der roten Liste der bedrohten Arten der Weltnaturschutzunion. Dieses scheue Raubtier ist mit seinen unbehaarten Lippen, der schmalen Zunge und den schließbaren Nasenlöchern extrem gut an seine Ernährungsgewohnheiten (überwiegend Termiten) angepasst. In einigen Regionen Indiens werden die Jungtiere gefangen und als Tanzbären eingesetzt. Hauptsächlich werden sie jedoch durch die Zerstörung

Das Damenkränzchen der Sambarhirsche wittert eine Gefahr. Der aufgestellte Schwanz ist ein deutliches Merkmal.

This sambar senses danger. The erect tail sends a clear signal.

Bandhavgarh

se closing times, when tourists have to leave the parks, are very strict, as they were at all the parks we visited. Drivers and guides who do not comply with these times can be banned from working for up to six months. Fortunately, India is also a country of storytellers, so shrewd guides still manage to come up with a legendary tale or two to make up for a dusty, break-neck race to get out of the parks on time.

Encounters with various species of deer, such as muntjacs, nilgai (blue bull antelope) or chitals (spotted deer), as well as rhesus macaques and langurs, are virtually guaranteed. Sloth bears, Old World porcupines and sambar deer also live hidden in the forest, while hyenas, foxes and jackals are occasionally to be seen on open ground. The park is also home to around

Bandhavgarh National Park	
State	Madhya Pradesh
Nearest city	Umaria
Size [sq mi]	173
Protected since	
National Park since	1968
Best time to travel	Mid-October – early June
Zones	5
Landscape	Sal trees, mixed forest, grassland
Fauna	
Predators	Bengal tiger, leopard, sloth bear, golden jackal, striped hyena, mongoose, jungle cat, wild dog
Grazing animals	Muntjac (barking deer), sambar deer, chital (spotted deer), gaur (Indian bison), wild boar, Old World porcupine, four-horned antelope, Indian pangolin
Birds	Red junglefowl, white-throated woodpecker, long-billed vulture, stork, fish owl, hornbill, eagle, falcon, Old World flycatcher
Primaten	Rhesus macaque

200 bird species. The gaur, too, has successfully been reintroduced in the park, having inexplicably disappeared at the end of the 1990s. The gaur is the largest living member of the bovine family. A bull can reach an impressive head-to-rump length of close to 11 ft, a height of just over 7 ft at the shoulder, and can weigh over 2,200 lbs. Listed as vulnerable, gaur live in hilly areas and can climb to altitudes of around 6,000 ft. They tend to have a mixed diet of grasses, leaves and herbs.

Among the mammals in the park, along with the tiger, the sloth bear is also on the International Union for Conservation of Nature's red list of threatened species. With its hairless lips, narrow tongue and closable nostrils, this shy predator is ext-

Bandhavgarh

ihres Lebensraums, durch Waldrodungen und durch die Einebnung von Termitenhügeln, in ihrem Bestand bedroht. Der Hindi-Name des Bären, Bhalu, inspirierte Rudyard Kipling offensichtlich zu seinem uns allen bekannten Charakter Baloo der Bär im Dschungelbuch.

Es gibt zwei wichtige Safaribegriffe, die jeder Guide beherrscht und jeder Besucher dringend beachten sollte. Der erste heißt: „Alarmcall!". Auf jeder unserer Fahrten machten die Guides uns auf die Rufe aufmerksam. Keiner von uns hätte sie von der „normalen" Kommunikation der Tiere untereinander unterscheiden können. Weithin hörbar kündet der Ruf der Hirsche und Affen und manchmal auch der Pfauen vom Standort und vom zurückgelegten Weg der gesuchten Raubkatze. Denn wo immer ein Tiger auftaucht, sein Erscheinen erzeugt erhöhte Alarmbereitschaft bei den Tieren und angespannte Aufmerksamkeit bei den Touristen. Sobald jedoch eine der Großkatzen tatsächlich erscheint, kommt der zweite Ausruf zum Einsatz: „Tiger! Tiger!". Und plötzlich bricht betriebsame Hektik aus. Beutetiere ergreifen die Flucht, Touristen ergreifen voller Aufregung ihre Kameras. Da gibt es kein Überlegen mehr, ob jetzt das richtige Objektiv aufgesetzt oder die Kamera entsprechend eingestellt ist. Der Blick in die Augen eines Tigers zwingt einen in die Knie und lässt jeden bis ins Innerste erschauern.

Hoch konzentriert halten die Guides und Fahrer ständig nach den Großkatzen Ausschau. Die meist sehr sandigen Pisten

Zuerst künden weithin hörbare Rufe der Hirsche von der Gegenwart des Königs des Dschungels – und mit dem nötigen Quentchen Glück kommt er dann auch in Fotoreichweite.

Loud deer calls announce the presence of the king of the jungle – and with a little luck, he also comes close enough for a photo.

remely well adapted to its dietary habits (it primarily dines on termites). In some regions of India, young sloth bears are caught and kept as dancing bears. Mostly, though, they are threatened by the destruction of their natural habitat, deforestation and the levelling of termite mounds. The Hindi name for this bear, Bhalu, was seemingly Rudyard Kipling's inspiration for naming his famous character Baloo the bear in „The Jungle Book" .

There are two essential safari calls every guide knows and every tourist should listen out for. The first is „Alarm call!" Our guides pointed these calls out to us on every single drive. None of us could have told the difference between these calls and how the animals „normally" communicate with each other. Audible over great distances, the call of deer and monkeys, and sometimes peafowl, announces the location and path taken by the big cats. Whenever a tiger is in the vicinity, its presence always puts the other animals on alert and attracts the undivided attention of tourists. But as soon as one of these big cats does actually appear, we hear the second call: „Tiger! Tiger!" And suddenly everyone's on the move. The tiger's likely prey flees, tourists excitedly grab their cameras – no time to worry about whether the right lens is on or if the settings are right. One look into the eyes of a tiger is enough to make even the seasoned observer go weak at the knees and tremble to the core.

Highly concentrated, guides and drivers are constantly on the look-out for big cats. It is a big help that the park's tracks are mostly sandy, as any movement by game cats leaves easily identifiable paw prints. The more often we are out in the park, the more frequently our gaze falls to the ground in the hope of an encounter with the striped predator.

Wie andere Großkatzen besitzt auch der Tiger eine runde Pupille. Die Iris ist in der Regel gelb.

Like other big cats, the tiger also has circular pupils. The iris is usually yellow in colour.

Pat und Patachon in der Affenversion.

The Laurel and Hardy of the simian world.

Bandhavgarh

des Parks sind eine hilfreiche Unterstützung. Jeder Wildwechsel hinterlässt leicht erkennbare „Tatzen"-Spuren. Je öfter wir unterwegs sind, um so häufiger wandern auch unsere Blick auf den Boden, immer in der Hoffnung auf eine Begegnung mit dem gestreiften Jäger.

Angefangen hat in Bandhavgarh alles mit Sita und Charger, einem legendären Tigerpaar. Sita, auch die Mutter von Bandhavgarh genannt, war die wohl am meisten fotografierte Tigerin der Welt. Ihr Porträt schaffte es sogar bis auf die Titelseite des National Geographic Magazins. Mit Charger, dem ersten gesunden Tigermännchen in den 90er-Jahren in Bandhavgarh, hatte sie vier Würfe von Jungtieren, bevor sie im Jahr 1996 auf rätselhafte Weise verschwand. Bis heute ist nicht definitiv geklärt, ob sie Opfer von Wilderern wurde oder eines natürlichen Todes gestorben ist. Für Charger, der seinen Namen aufgrund seines extrem aggressiven Verhaltens gegenüber Safarijeeps und Elefanten erhielt, wurde dieses Ereignis zum tragischen Wendepunkt in seinem Leben. Vier Jahre später unterlag der damals ca. 20-jährige Tiger nach mehr als einem Jahrzehnt in einem Revierkampf seinem Herausforderer. Sein Körper wurde am „Charger-Point" begraben, wo heute noch eine Gedenktafel an ihn erinnert.

Eine erfolgreiche Tigersichtung hängt nicht nur vom Wissen und Geschick des Guides und des Fahrers ab. Mancher holt sich dafür auch noch die Unterstützung indischer Götter. Speziell Ganesha wird angerufen, jeden Morgen mit einem hoffungsvollen Streicheln über den Rüssel, wenn man Glück und Erfolg am Beginn einer Unternehmung braucht. Fast jedes Hotel oder jede Lodge hat im Eingangsbereich eine Statue dieses Elefantenkopf tragenden Götterboten stehen, die diesen „göttlichen Beistand" ermöglicht.

Tiger leben in der Regel als Einzelgänger.

Tigers are generally loners.

Zur Abkühlung und zum Trinken kommen die Tiger an den Fluss.

Tigers head to the river to cool down and drink.

Everything began in Bandhavgarh with Sita and Charger, a legendary tiger coupling. Sita, also known as the Mother of Bandhavgarh, was considered the most photographed tiger in the world. She even made it onto the cover of National Geographic. She had four litters sired by Charger, the first healthy male tiger in Bandhavgarh in the 1990s, before she disappeared in mysterious circumstances in 1996. To this day, it is not known for sure whether she was killed by poachers or whether she died a natural death. Her death marked a tragic turning point in the life of Charger, named for his extremely aggressive behaviour towards safari vehicles and elephants. Four years later, at the age of 20, he succumbed to a challenger in a territorial fight after more than a decade as the top cat in the park. He was buried at Charger Point, where there is now a plaque in his honour.

A successful tiger sighting is not only down to the knowledge and skill of the guide and the driver; some also call upon the assistance of Indian gods. Ganesha in particular is invoked every morning with a hopeful stroke along the trunk when seeking luck and success at the beginning of an undertaking. Almost every hotel and lodge has a statue of this elephant-headed deity at its entrance to offer staff and guests alike divine assistance.

Autor/Author: Kerstin von Splényi

Die weißen Flecken auf der Ohrenrückseite, die Augenflecken, gaukeln einem eventuellen Angreifer vor, dass er beobachtet wird.

The white marks on the back of the ears look like eyes and fool any would-be attackers into thinking they are being watched.

Der Tiger ist die größte aller lebenden Katzenarten und aufgrund des charakteristischen dunklen Streifenmusters auf goldgelbem bis rotbraunem Grund unverwechselbar.

The tiger is the largest of all living cats and its characteristic dark stripes on a yellow-gold to reddish-brown background make it unmistakeable.

Die Herde Axishirsche (Mitte) löscht noch ihren Durst kurz vor Sonnenuntergang und macht sich, wie auch die Rotte Wildschweine (oben), auf den Heimweg. Auf unserem Heimweg fanden wir noch diese farbenprächtige Eidechse (unten).

A herd of chitals (middle) quenches its thirst before sunset and, like the sounder of wild boar (top), heads home. We also head back, seeing this vibrantly coloured lizard on our way (bottom).

Wenn es lautstarke Zwistigkeiten gibt (oben), ruft das die aufmerksamen Mungos auf ihren Wachposten (unten).

Loud disputes (top) bring the watchful mongoose to their look-out posts (bottom).

Bei mehr als 40 Grad Hitze treibt es alle an die Wasserlöcher, Affen genauso wie Axishirsche und Elefanten.

Temperatures soar to over 40 degree centigrade and force simians, chitals and elephants alike to the watering holes.

Der farbenprächtige Pfau (links) und der Wellenbrust-Fischuhu (rechts) gehören zu den beeindruckenden Vögeln in Bandhavgarh.

The colourful Indian peafowl (left) and the brown fish owl (right) are among Bandhavgarh's impressive array of birdlife.

Auch bei der Affenbande gibt es Streit. Die Mutter mit Jungtier beäugt das Ganze misstrauisch.

The simians have their disputes, too. This mother protects her young while warily observing.

Rhesusaffen (oben) und Languren (unten) sind in Indiens Nationalparks die häufigsten Affenarten.

Rhesus macaques (top) and langurs (bottom) are the most common simians found in India's national parks.

Bengalische Tiger

The Bengal tiger

Bengalische Tiger sind nach dem Sibirischen Tiger die größten Raubkatzen der Erde. Sie werden auch Königstiger oder Indische Tiger genannt. Einst waren sie von Südosteuropa bis in den Mittleren und Nahen Osten beheimatet. Heute leben die letzten freilebenden ca. 2.100 Tiere in verschiedenen Nationalparks in Indien. Hier werden sie streng geschützt und Touristen haben für einige Wochen im Jahr die Möglichkeit, die imposanten Großkatzen wild und natürlich zu erleben. Bis 1930 lebten 40.000 Tiger in Britisch-Indien – durch die organisierte Tigerjagd wurden mehr als 30.000 Exemplare in nur drei Jahrzehnten getötet. Von ursprünglich neun Tiger-Unterarten sind drei bereits ausgestorben und alle anderen extrem bedroht.

The Bengal tiger is the second largest of the big cats after the Siberian tiger. Also known as royal tigers or Indian tigers, they once roamed an area from south-eastern Europe to the Middle East. Today, the last 2,100 or so of their kind outside captivity live in various Indian national parks, where they are closely protected and tourists have the opportunity to see these impressive big cats in their natural habitat during a few weeks a year only. Until 1930, 40,000 tigers lived in British India. More than 30,000 were then killed in only three decades by organised big game hunting. Of the original nine sub-species of tiger, three are already extinct and all others are listed as critically endangered.

Bengalische Tiger | The Bengal tiger

Ein stattlicher Tiger posiert geradezu für die Fotografen ...

A stately tiger poses for the photographers ...

Tiger sind optimal getarnt und vor dem Dschungelhintergrund sowie in der Bodenvegetation nahezu unsichtbar, ihre Streifenzeichnung ist im Tierreich unverwechselbar. Nur wenn sie sich bewegen, sind sie aus kurzer Entfernung von der Landschaft zu unterscheiden. Schwarze, breite Querstreifen ziehen sich vom Kopf bis zum Schwanz und sind bei jedem Tier individuell. Acht bis neun, meist doppelte, schwarze Ringe zieren den Schwanz. Tiger haben schwarze Ohren mit einer auffälligen weißen Markierung.

Männliche Bengalische Tiger werden bis zu 300 Kilo schwer, bei einer Körperlänge von 270-310 cm. Weibchen können bis 180 kg schwer werden. Die Schulterhöhe erreicht 90-100 cm. Tiger sind Einzelgänger und bevorzugen als Lebensraum dichte Vegetation und Wassernähe. Sie können gut schwimmen und kühlen sich gerne im Wasser ab. Männchen beanspruchen zwei bis sieben Weibchen in ihrem Revier, die jeweils in getrennten Gebieten leben und dulden dort keinen Rivalen. Sie markieren mit Urin und zusätzlich mit Kratzspuren an Bäumen regelmäßig ihr Revier. Ein Tigerpaar findet sich meist nur für zwei Tage zusammen und sie vollziehen die jeweils kurze Paarung bis zu 52 Mal pro Tag.

Damit ausgewachsene Tiger satt werden, brauchen sie mindestens 8 Kilo Fleisch am Tag. Sie jagen hauptsächlich große Säugetiere wie Rinder, Rehwild oder andere Huftiere sowie Wildschweine und ab und zu auch kleinere Säugetiere wie Affen und Kaninchen. Tiger schleichen sich meist gegen den Wind bis auf zehn Meter an ihre Beute heran oder lauern ihr in der Dämmerung oder Dunkelheit auf. Sie springen bis zu sechs Meter weit, die Gejagten können nur durch Schnelligkeit entkommen, denn Tiger brechen eine Verfolgung meist schon nach 100 bis 200 Metern ab. Beim Verzehr der Beute

Bengalische Tiger | The Bengal tiger

Tigers are perfectly camouflaged and virtually invisible against a jungle background or hiding in ground cover. Their striped markings are unmistakeable in the animal kingdom; only when they move do they stand out against the landscape when seen from a short distance. Covered from head to tail with broad, black vertical stripes, each tiger has unique markings. The tail has eight or nine black, mostly double, rings. Tigers have black ears with a conspicuous white mark.

Male Bengal tigers can weigh up to 660 lbs and grow to 9 or 10 ft in length; females weigh up to 400 lbs. They can stand over 3 ft tall at the shoulder. Tigers are solitary animals and prefer to live in thick vegetation and close to water. They are strong swimmers and like to cool down in the water. Males claim two to seven females on their territory, each living in separate „home ranges", and do not tolerate any rivals. They regularly mark their territory with urine and scratches on trees. A tiger couple generally only stays together for two days and in that time they copulate up to 52 times a day.

Fully grown tigers eat at least 18 lbs of meat per day. They mainly hunt large mammals such as cattle, deer or other

Eine Tigerin mit ihren Jungen: Im Zweifel gehört ihnen auch die Straße!

A female tiger with her young occupies the road ahead.

Bengalische Tiger | The Bengal tiger

trinken Tiger sehr viel und bevorzugen einen Riss in Wassernähe.

Ein Tigerweibchen kann zwei bis sechs Junge gebären. Tigerbabys kommen blind zur Welt und können erst nach rund zwei Wochen sehen. Danach verlassen sie mit ihrer Mutter das sichere Versteck. Nach 18 Monaten verlieren junge Tiger ihre Milchzähne, erst dann lernen sie die Jagdtechniken und bleiben noch bis zu drei Jahre lang bei ihrer Mutter, bevor sie sich ein eigenes Revier suchen müssen. Die Sterblichkeit junger Tiger ist sehr hoch – ein Weibchen kann in seinem Leben oft nur vier bis fünf Jungtiere bis zur Selbstständigkeit aufziehen. Tiger erreichen mit fünf Jahren ihre größte Kraft und Geschicklichkeit und werden in freier Natur etwa 18 Jahre alt.

Erst seit 1970 sind das Töten und der Tigerfellhandel verboten. In den letzten zwei Jahrzehnten haben die Tiger trotzdem weitere 40 % ihres Lebensraums durch die starke Ausbreitung des Menschen verloren, ihre Existenz im Dschungel ist akut gefährdet.

Eleganz und Wachsamkeit – eng verbunden.

Elegance and watchfulness in perfect harmony.

Bengalische Tiger | The Bengal tiger

hoofed animals, such as wild boar, and occasionally smaller mammals such as monkeys and rabbits. Tigers usually sneak up on their prey from downwind, approaching to a distance of around ten yards, or stalk it at dusk or in darkness. They can jump up to six yards and their prey's only hope of escape is to use sheer speed because tigers generally break off the chase after 100 to 200 yards. Tigers drink a great deal when consuming their prey and therefore prefer to kill close to water.

A female tiger has between two and six cubs in a single litter. Tiger cubs are born blind and are only able to see after around two weeks. After that, they leave the safety of cover with their mother. Juvenile tigers lose their milk teeth after 18 months and only then do they begin to learn to hunt, staying with their mother until they are three years old, when they have to seek out their own territory. The mortality rate of juvenile tigers is very high; a female often only manages to raise four or five cubs to maturity in her lifetime. Tigers are at their strongest and most agile at the age of five and live to around 18 years of age in the wild.

The killing of tigers and trade in tiger skins has only been illegal since 1970. Nevertheless, tigers have lost a further 40 per cent of their natural habitat over the last two decades due to the spread of humans, and their survival in the jungle is critically endangered.

Autor/Author: Harry P. Lux

Dieser Tiger möchte nicht gestört werden! Auch nicht für Fotos ...

Do not disturb! Not even for photos ...

Asiatische Löwen

The Asiatic lion

Löwen lebten schon vor 3,5 Millionen Jahren in Afrika. Die Entwicklung der Unterart, die heute noch in Gir lebt, geschah vor etwa 50.000 bis 100.000 Jahren. Die Mähne der Asiatischen Löwenmännchen ist sichtbar weniger stark ausgeprägt und im Bereich der Ohren schwach entwickelt. Die Löwen in Gir haben eine längere Behaarung an den Ellenbogen und die schwarze Schwanzspitze ist oft buschiger. Asiatische Löwen haben fast immer eine Bauchfalte, die längsseits in der Bauchmitte verläuft. Löwenmännchen werden bis zu 190 kg schwer bei einer Körperlänge von 290 cm. Die Schulterhöhe erreicht 110 cm.

Lions have been living in Africa for around 3.5 million years. The sub-species that still lives in Gir developed between 50,000 and 100,000 years ago. The mane of the male Asiatic lion is visibly less pronounced and thin around the ears. The lions in Gir have longer hair on their elbows and the tuft of black hair at the tip of the tail is often bushier than that of their African counterparts. Asiatic lions almost always have a longitudinal fold of skin, running the length of their body, along the middle of the belly. Adult males can weigh up to 420 lbs and grow to a length of 9 ½ ft, typically reaching around 3 ½ ft at the shoulder.

Asiatische Löwen | The Asiatic lion

Auch Raubkatzen sind untereinander verschmust. Der Bruder beäugt das Ganze aufmerksam.

Even big cats like to nuzzle. The brother looks on attentively.

Der bevorzugte Lebensraum Asiatischer Löwen sind Laubwälder, Savannen mit Dornwäldern und trockene Gebiete. Die Rudelstärke beträgt oft nur zwei bis sechs Tiere, deutlich kleiner als bei den Afrikanischen Löwen mit bis zu 30 Tieren. Junge Löwen werden nach sechs Monaten von der Mutter entwöhnt und bleiben dann noch rund zwei Jahre im Rudel und lernen die Jagd. Zu ihrer bevorzugten Beute gehören Sambar- und Axis-Hirsche, Antilopen, Gazellen und Wildschweine. Löwen müssen in die Nähe ihrer Beute kommen oder sie einkreisen, da sie eine Geschwindigkeit von bis zu 65 km/h nur kurz durchhalten können. Im Rudel fressen männliche Löwen zuerst, dann die ranghöchsten Löwinnen und zuletzt die jungen Löwen. In Indien wurden die Löwen Anfang des 20. Jahrhunderts praktisch ausgerottet. Mit nur 20 Exemplaren und in Zoos gehaltenen Tieren konnte sich unter strengem Schutz wieder eine kleine Population aufbauen. Der Asiatische Löwe ist durch Inzucht bedroht, der Gen-Pool der Tiere ist sehr begrenzt und es leben zur Zeit nur etwa 175 zeugungsfähige Exemplare. Löwen können sich ganzjährig fortpflanzen. Rund 110 Tage später kommen zwei bis vier hilflose Junge zur Welt. Nach vier Jahren sind die Löwinnen geschlechtsreif, die Männchen folgen ein Jahr verzögert. Sie können sich bis zum 15. Lebensjahr fortpflanzen. Weibchen werden 18-20 Jahre alt, etwa zwei Jahre älter als die Männchen. Die Asiatischen Löwen gelten als extrem bedroht. Im letzten Jahr wurden 100 Junge im Gir Nationalpark geboren, dass sind 20 mehr als sonst

Asiatische Löwen | The Asiatic lion

The preferred habitat of the Asiatic lion is deciduous forests, savannah with thorny forests, and dry areas. They often live in prides of only two to six lions, far fewer than African lion prides, which can number up to 30. Juvenile lions are generally weaned by the age of six months and remain with the pride for around two more years, learning to hunt. Their preferred prey includes sambar and chitals, antelopes, gazelles and wild boar. Lions have to get close to their prey or encircle them because, although they can reach a top speed of up to 40 mph, they can only maintain it for short periods. Within a pride, the males feed first, followed by the highest ranking females and finally the juvenile lions. Asiatic lions were practically hunted to extinction in India in the early 20th century. With only 20 lions left, plus those living in captivity, a small population was able to grow again under close protection. Asiatic lions are threatened by inbreeding as their gene pool is very limited and there are currently only around 175 lions capable of breeding. Lions can breed all year round and give birth to two to four helpless cubs following a gestation period of around 110 days. Female lions are sexually mature after four years, males one year later, and can breed until the age of 15. Females live to an age of 18 to 20 years, two years longer than males. Asiatic lions are listed as critically endangered. 100 cubs were born in the Gir Forest National Park last year, 20 more than usual.

Autor/Author: Harry P. Lux

Einen wachen klaren Blick wird der Kleine brauchen, um zu überleben.

This young lion will need to stay alert and watchful if he is to survive.

Panzernashörner

The Indian rhinoceros

Panzernashörner (auch indische Nashörner genannt) waren früher über den gesamten indischen Subkontinent verbreitet, ihre Vorgänger noch weiter – heute leben nur noch ungefähr 2.800 Tiere im Norden Indiens und in Nepal in geschützten Gebieten – davon etwa 70 Prozent im Kaziranga Nationalpark. Sie sind stark gefährdet und vom Aussterben bedroht. Neben ihrem afrikanischen Vetter, dem Breitmaulnashorn, und drei Arten von Elefanten, sind sie die fünftgrößten Landsäugetiere der Welt. Ihren deutschen Namen haben sie von ihrer etwa vier Zentimeter dicken Haut, die besonders am oberen Ende der Beine und im Gesäßbereich dicke Falten bildet. Ihr einzelnes Horn sitzt auf der Nase, kann bis zu 30 cm lang werden und etwa zwei Kilo wiegen.

In earlier times, the Indian rhinoceros, also called the greater one-horned rhinoceros, ranged across the entire Indian sub-continent, and its predecessors were spread even further. Today, there are only 2,800 or so rhinos left living in conservation areas in northern India and Nepal, around 70 per cent of them in Kaziranga National Park. They are critically endangered and at risk of extinction. The Indian rhinoceros is the fifth largest land mammal after the three species of elephant and its African cousin, the white (or square-lipped) rhinoceros. The skin of the Indian rhinoceros is approximately 1.5 inches thick and forms deep folds, particularly at the top of the legs and around the rump. The single horn grows up to a foot long on the animal's nose and can weigh up to 4 ½ lbs.

Bereits bei den Kälbern sind die Hautfalten am Hals stark ausgeprägt.

The folds of skin on the neck are already clearly visible on rhino calves.

Imposante Panzernashorn-Bullen erreichen bei einer Körperlänge von 3,70 bis 3,80 m und einer Schulterhöhe von 1,70 bis 1,90 m ein Gewicht von 2,2 Tonnen und mehr ... und das von „nur" rund 150 Kilo Gras pro Tag! Ihre Hörner nutzen sie weniger als Waffe, da sind die dolchartigen, schräg nach vorne stehenden unteren Schneidezähne eher geeignet, einem Gegner gefährliche Wunden beizubringen. Eine Behaarung fehlt ihnen fast völlig, nur an den Wimpern, Ohren und am Schwanz sind wenige Haare vorhanden.

Kühe werden mit etwa fünf bis sieben Jahren geschlechtsreif, männliche Tiere im Alter von etwa acht bis zehn Jahren. Eine Paarung ist das ganze Jahr über möglich, danach trennen sich die Partner wieder und etwa 16 Monate später wird ein einzelnes Kalb geboren. Dieses wiegt bei der Geburt bereits bis zu 75 Kilo und kann nach 30 Minuten schon alleine stehen! In den ersten Tagen trinkt es bis zu 30 Liter Milch und nimmt täglich knapp zwei Kilo zu. Kälber bleiben etwa drei Jahre bei der Mutter, bevor sie vor einer neuen Geburt vertrieben werden.

Panzernashörner haben in freier Natur eine Lebenserwartung von ca. 40 Jahren. Sie haben kaum natürliche Feinde, es kommt allerdings vor, dass Tiger manchmal ein Kalb reißen. Die größte Gefährdung bringt der Mensch, der einerseits den Lebensraum der Tiere immer weiter einengt und sie andererseits wegen ihres Horns jagt, dem in der asiatischen Medizin immer noch heilende Wirkungen nachgesagt werden.

Bevorzugter Lebensraum der Panzernashörner sind offene Graslandschaften mit kurzen Gräsern und sumpfigen Anteilen mit hohem Elefantengras, das als Versteck dienen kann. Kleinere lichte Wälder sind ebenfalls zur Abwechslung willkommen. Dabei profitieren die Nashörner von den Elefanten, die immer wieder dichte Waldfronten aufreißen. Panzernashörner hören sehr gut und sehen sehr schlecht. Sie sind sehr gute Schwimmer und verbringen viel Zeit im Wasser. Sie können unter Wasser grasen. Die meiste Zeit des Tages suhlen und wälzen sie sich am liebsten in Schlammlöchern. Das ist gut für die Kühlung und gegen Parasiten

Panzernashörner | The Indian rhinoceros

A fully grown male Indian rhinoceros can grow to 12 to 12 ½ ft long, reach a height of 5 ½ to 6 ¼ ft at the shoulder, and weigh 4,850 lbs – all on „just" 300 lbs or so of grass a day! They tend not to use their horn as a weapon; their dagger-like lower incisors, which are angled forward, are better suited to inflict serious wounds on an opponent. They have very little hair, aside from eyelashes, a little hair around the ears, and a tail brush.

Females reach sexual maturity at around five to seven years of age, males at between eight and ten. They can breed at any time of year, after which the partners go their separate ways and a single calf is born 16 months later. It already weighs up to 165 lbs at birth and is capable of standing without assistance after only 30 minutes. It drinks up to 50 pints of milk in the first days of its life and gains close to 4 ½ lbs in weight per day. Calves remain with their mother for around three years until they are driven away by the birth of the next offspring.

Indian rhinoceroses living in the wild have a life expectancy of around 40 years. They have barely any natural enemies, although tigers have been known to snatch a calf. The greatest danger is from humans, who are both encroaching ever more on their habitat and hunting them for their horns – in Asian medicine their horns are still believed to have healing properties.

The preferred habitat of the Indian rhinoceros is open grassland with short grasses and swampy areas with long elephant grass to hide in. Small, sparse woodlands are also a welcome change, and the rhinos benefit from the behaviour of elephants, who regularly rip out the dense edges of forests. Indian rhinos have excellent hearing but poor eyesight. They are strong swimmers, spend a great deal of time in the water and are able to graze under water. They prefer to spend most of the day wallowing and rolling in mud holes, which helps them to cool down and keeps away parasites.

Autor/Author: Harry P. Lux

Ob großes oder kleines Horn – diese Tiere strahlen eine urwüchsige Kraft aus!

Horn size isn't everything – these animals exude raw power regardless.

Hilfsprojekt | Aid project

Das Nepal-Tiger-Projekt …

… wurde im April 2011 ins Leben gerufen, als der Gründer John Brooks – ein pensionierter amerikanischer Bundesbeamter und Wildbiologe – zusammen mit Anna Bach, Fachfrau für Tibetanische Medizin, nach Nepal reiste, um sie bei ihrer Suche nach einer bestimmten Baumart zu unterstützen. John und Anna wanderten zwei Tage lang im Himalaya von einer entlegenen Bushaltestelle in den nepalesischen Okhaldhunga Distrikt. Dorfbewohner erzählten ihnen beiläufig, dass im Dschungel, der ihr Dorf umgibt, Tiger leben. Sie hatten Beweise dafür, dass sich innerhalb der letzten zwei Jahre mehrere Exemplare dieser seltenen Raubtiere in der Gegend aufgehalten hatten. Diese Informationen haben sie an führende Naturschutz-Organisationen weitergeleitet – es wurde jedoch nichts unternommen, um die Behauptungen nachzuverfolgen.

Im Herbst 2012 kehrte John mit dem Ziel nach Nepal zurück, das Vorkommen von Tigern im Okhaldhunga Distrikt zu bestätigen. Dorfbewohner wurden interviewt, „Kamerafallen" aufgestellt, und Kot-Spuren sowie Tatzenabdrücke dokumentiert.

Die gesammelten Beweise erhärteten die Vermutung, dass in der Gegend Tiger leben. Obwohl bisher keine Tiger in die Kamerafallen „tappten", konnte das Team Augenzeugenberichte notieren und freute sich darauf, zurückzukehren – in der Hoffnung, schlüssige Beweise für einen kleinen Tigerbestand zu finden. Wie ein Dorfbewohner formulierte: „Der Tiger kann dich sehen, und er wird sich zeigen, wenn er soweit ist." John ist davon überzeugt, dass es nur eine Frage von Zeit und Ausdauer ist, ein Tigerfoto zu bekommen.

Das Nepal-Tiger-Projekt suchte im September 2013 nochmals nach Beweisen. Es gab jedoch zu heftige Regenfälle, sodass das Team keine Spuren der Großkatzen finden konnte. Im Frühjahr 2014 spürte John mit einigen freiwilligen Helfern Leoparden, Schwarzbären, Languren und Kleine Pandas in dieser Gegend auf. Nachts wurden Tigerrufe gehört.

Im Juli 2014 traf sich Michael Matschuck, Vorstandsmitglied der Aktionsgemeinschaft Artenschutz (AGA) e.V., mit John Brooks, um eine Zusammenarbeit zu besprechen. Schnell wurde klar, dass die Idee und die Art dieses Projektes etwas Besonderes sind und die AGA dieses Projekt gern mit unterstützen wird. Einen wertvollen Beitrag werden die Erlöse aus dem Verkauf dieses Buches darstellen. Das Projekt arbeitet auch mit der einheimischen Dorfbevölkerung und dortigen Schulen zusammen, um die Menschen vor Ort für die Erhaltung der Tierwelt zu sensibilisieren, und ihnen Techniken zu vermitteln, die das Zusammenleben mit den Raubkatzen ermöglichen.

Hilfsprojekt | Aid project

The Nepal Tiger Project ...

... began in April 2011 when founder John Brooks, a retired federal agent and wildlife biologist, travelled to Nepal with Anna Bach, a practitioner of Tibetan medicine, to support her in the effort to locate a particular species of tree. John and Anna walked two days from a remote bus stop into the Okhaldhunga District of Nepal's Himalaya mountain range. Local villagers mentioned in passing that tigers were living in the jungle around their village. They had evidence that several of these reclusive predators had been in the area during the previous two years. This information was passed on to major conservation groups; however, no action was taken on their part to investigate the claims.

In autumn 2012, John returned to Nepal with the goal of confirming the presence of tigers in the Okhaldhunga District. Villagers were interviewed, camera „traps" were set, and scat samples and paw prints were documented. The evidence collected strengthened the case for the presence of tigers in the area. Although the tigers eluded the camera traps, the team was very excited to record eyewitness accounts of their presence and looked forward to returning on location to obtain conclusive proof of a small tiger population. As one villager put it: „The tiger can see you and will show himself when he is ready". John believes that capturing the tiger on camera is just a matter of time and persistence.

The Nepal Tiger Project continued looking for signs of tigers in September 2013. The rains were too heavy, however, and

did not allow the team to track the movement of large cats. In the spring of 2014, John and a few volunteers tracked leopards, black bear, langur monkeys, and red pandas in the area. Tigers' calls were heard at night.

Michael Matschuck, member of the board of Aktionsgemeinschaft Artenschutz, and John Brooks met in July 2014 to discuss a potential collaboration. It quickly became clear that this is a very special and interesting project and thus AGA is happy to support it in the future. The proceeds from the sale of the book will be a valuable contribution. The project also works together with local villagers and schools to raise awareness for the conservation of animals and their natural habitat, and to teach ways of living side by side with big cats.

John Brooks unterwegs in Tibet auf der Suche nach Tigern.

John Brooks searching for tigers in Tibet.

Aktionsgemeinschaft Artenschutz e.V.
Action Campaign for Endangered Species (ACES)

Liebe Leserinnen und Leser,

Naturfotografen sind eine besondere Spezies ... bei jedem Wetter laufen sie schwer bepackt mit Kameras, Objektiven und Stativ durch die Natur, auf der Suche nach dem einen Foto, das noch keiner hat – oder einem Bild, das sich von der breiten Masse abhebt. Dank moderner Digitalfotografie werden dabei heute viele Tausend Bilder in kurzer Zeit gemacht. Mich packte die Naturfotografie auf meiner ersten Fernreise. Eine fremde Welt tat sich auf und ich konnte die Kamera kaum weglegen, um die Eindrücke auch zu Hause meiner Familie und Freunden zeigen zu können. Es war um mich geschehen. Ich flog in der Welt herum, um zu fotografieren. Meine Reisen wurden immer ausgefallener und schon bald konnten Pauschalreiseanbieter meine Wünsche nicht mehr erfüllen. Also begann ich, sie selbst zu organisieren. Dabei lag der Fokus immer darauf, Tiere in freier Natur vor die Kamera zu bekommen.

Aus dem Hobby ist mein Beruf geworden und heute biete ich als Reiseveranstalter Natur- und Fotoreisen in die ganze Welt an und begleite diese so oft mir das möglich ist. Die Kamera ist mein ständiger Begleiter und es bieten sich immer wieder neue Möglichkeiten, Erlebnisse und Situationen in der Natur festzuhalten.

Vor zwei Jahren nutzte ich die Chance, einige meiner Bilder im Fotografeneditionen-Bildband „Mittel- und Südamerika" zu veröffentlichen und damit einem breiten Publikum zugänglich zu machen. Für jeden passionierten Naturfotografen ist der Abdruck seiner Bilder in einem Buch ein absolutes Highlight. Daher freut es mich nun, auch bei diesem Buch mitwirken zu können. Es freut mich besonders, dass ich das Projekt als Reiseveranstalter unterstützen durfte und so den Fotografen die Möglichkeit geben konnte, die schönen Bilder dieses Buchs aufnehmen zu können.

Indien ist eine Reise in eine andere Welt. Die Nationalparks begeistern jedes Jahr viele Besucher. Ihre Fauna und Flora sind so unterschiedlich, dass es ratsam ist, mehrere Nationalparks zu besuchen. Tiger findet man mit etwas Glück in fast allen Nationalparks Indiens. Wer hingegen Panzernashörner, wilde Elefanten oder die letzten Löwen Asiens sehen möchte, der muss seine Reise schon sehr genau planen. Durch unsere langjährige Erfahrung ist es uns möglich, für Sie ein individuelles und auf ihre Wünsche hin ausgearbeitetes Angebot zu erstellen. Bei Indien dürfen auch Kultur und Menschen nicht vergessen werden, es ist ein vielfältiges Land und wer es zum ersten Mal bereist, merkt schnell, dass es nicht seine letzte Reise in diese fremde Welt war. Infos und Kontaktdaten zu unseren Reiseangeboten finden Sie unter **www.eaglerayreisen.com.**

Ihr Martin Ruhnke

Reiseveranstalter | trip organiser

Dear reader,

Nature photographers are different from the rest of us ... Weighed down with all sorts of cameras, lenses and tripods, they make their way through the great outdoors in search of that unique photograph – one that stands out from the rest. These days, digital photography makes it possible to take thousands of shots in a very short space of time – which also increases the overall quality of the photos taken. Nature photography worked its magic on me on my first big trip far from home. I discovered a whole new world and could hardly put my camera down as I wanted to be able to share everything I had seen with my friends and family back home. That was it for me. Some people spend their money on fast cars – I flew round the world to take pictures. My trips became more and more exotic and soon package holidays just didn't meet my needs any more. And so I began to organise my own trips, with the focus always on capturing pictures of animals living in the wild.

I turned my hobby into my profession and now organise nature and photo holidays all over the world, going along myself whenever possible. My camera never leaves my side and there are always plenty of chances to record new experiences and situations.

Two years ago I took the opportunity to publish some of my pictures in the „Central and South America" Fotografeneditionen coffee-table book and make them available to a wider reading public. Seeing your photographs printed in a book is a particular highlight for any avid nature photographer. It is therefore a pleasure for me to be involved in the creation of this book: A photo book focusing on the nature and wildlife of India. I am particularly pleased to have been able to support this project by organising the trip, giving the photographers the opportunity to take the beautiful photos that make up this book.

Der Reiseveranstalter Martin Ruhnke mit der Unterwasser-Kamera auf Reisen.

Martin Ruhnke, trip organiser, with his underwater camera.

India is a journey into a different world. Its national parks enthral thousands upon thousands of visitors every year. Its fauna and flora are so diverse that it's advisable to visit several different national parks. Tigers are to be found in almost all of India's national parks, with a bit of luck. If you're looking for Indian rhinos, wild elephants or the last lions in Asia, however, you have to plan your trip much more carefully. With our many years of experience organising nature and photo holidays, we can put together an itinerary tailored to your wishes. A visit to India should not overlook the culture and the people; it is a very diverse country and anyone going there for the first time will quickly realise that it will not be their last trip to this unfamiliar land. For more information, contacts and details of the trips we offer, visit **www.eaglerayreisen.com.**

Best regards, Martin Ruhnke

Fotografen-Viten | Photographer profiles

Katrin und Jens Beck haben anlässlich ihrer Hochzeitsreise im Jahr 2007 nach Alaska ihre gemeinsame Faszination für Natur, Wildnis und Abenteuer entdeckt. Anlässlich dieser Reise hat Jens seine (seit seiner Jugend beendete Leidenschaft) für die Natur- und Tierfotografie wieder voll aufleben lassen und Katrin damit angesteckt. Seitdem sind die beiden jedes Jahr gemeinsam ein- bis zweimal mit ihrer Fotoausrüstung unterwegs gewesen – in der Natur Nordamerikas, der Arktis, Afrika, und jetzt im Rahmen dieses Buchprojekts erstmals in Indien.
Katrin wurde 1978 in Bautzen/Sachsen geboren und lebt seit zwölf Jahren in Radebeul. Nach ihrem Studium zur Diplomkauffrau an der HTW Dresden betreibt sie jetzt eine Filiale des „The Fresh Tea Shop" in Sachsens Hauptstadt.
Jens wurde 1966 in Hamburg geboren. Nach seinem Studium kam er im Sommer 1990 nach Dresden und etablierte dort sein bis heute bestehendes Maklerunternehmen in Radebeul. Zudem ist er seit 15 Jahren als öffentlich bestellter und vereidigter Sachverständiger für Immobilienbewertung tätig.
Neben Reisen und Tierfotografie ist der Triathlonsport eine große Leidenschaft der beiden. Sie leben in Radebeul und haben einen fünfjährigen Sohn.

Katrin and Jens Beck discovered their shared fascination with nature, the wilderness and adventure on their honeymoon to Alaska in 2007. It was on that trip that Jens rediscovered his youthful passion for photographing nature and animals, and Karin was soon hooked, too. Since then, they have travelled to distant countries once or twice a year, always taking their camera equipment with them, photographing the great outdoors in North America, the Arctic, Africa and now, for the first time as part of this book project, India.
Katrin was born in Bautzen, in Saxony, in 1978 and has lived in Radebeul for twelve years. Having graduated in business studies at Dresden University of Applied Sciences, she now runs a branch of The Fresh Tea Shop in Dresden, the capital city of Saxony.
Jens was born in Hamburg in 1966. After completing his studies he came to Dresden in the summer of 1990 and established his estate agency, which he still runs to this day, in Radebeul. He has also been a publicly appointed and sworn property valuer for 15 years. In addition to travelling and photographing animals, they both share a passion for competing in triathlons. They live in Radebeul and have a five-year-old son.

Frank Hanel, Jahrgang 1961, führt vom kleinen ostwestfälischen Ort Barntrup aus seine Firma, die sich auf Ersatzteilhandel für klassische Alfa Romeos spezialisiert hat. Wann immer der sportliche Westfale Zeit hat, ist er aber draußen. Mit dem Mountain-Bike unterwegs im Weserbergland, auf Klettertour in den Alpen oder auf einer Bergtour in Afrika. Immer mit dabei: seine Kamera. Naturfotografie ist die große Leidenschaft des 54-Jährigen. Viel Geduld müssen seine Frau und seine beiden Töchter deshalb mitbringen, wenn sie mit ihrem Vater auf Reisen gehen: Jede Tour mit ihm dauert mindestens doppelt so lang wie geplant, weil Frank noch hier eine seltene Spinne und dort eine schöne Blüte entdeckt, die er im richtigen Licht in Szene setzen muss. In Indien war er für das Fotoprojekt das erste – aber nicht das letzte Mal.

Frank Hanel, born in 1961, runs a business in the small Westphalian town of Barntrup specialising in spare parts for classic Alfa Romeos. He is outdoors whenever he can find the time, out on his mountain bike in the Weser Uplands, climbing in the Alps or on a tour of Africa's mountains. His camera is his constant companion. Nature photography is the 54-year-old's great passion, which demands a great deal of patience from his wife and their two daughters when they join him on his trips. With Frank, every tour takes at least twice as long as planned, because he is constantly discovering a rare spider here and a beautiful flower there, all of which has to be set in the right light. This trip to India for the photo project was his first, but certainly not his last.

Fotografen-Viten | Photographer profiles

Andreas Klotz wurde 1968 geboren. Nach einer Berufsausbildung zum Schriftsetzer hat er sich bereits mit 19 Jahren selbstständig gemacht. Heute ist er Inhaber und Geschäftsführer der TiPP 4 GmbH, Werbeagentur und Verlag in Rheinbach bei Bonn. Seit über 30 Jahren ist die Naturfotografie sein Hobby. Andreas ist verheiratet und Vater von zwei Teenagern. Ab und zu „darf" er auf Trekking- oder Foto-Reisen mal für zwei bis drei Wochen weg von seiner Familie, seinem Büro und der täglichen Routine. Dann zieht es ihn möglichst weit weg in die Natur, vor allem, um intensiv zu fotografieren und die Erlebnisse in Reiseberichten festzuhalten. Die Reiseleidenschaft packte ihn 1988 auf seiner ersten Fotosafari in Kenia. Er war mehr als zwanzigmal in Afrika, in Südamerika und für dieses Projekt erstmalig in Indien.

Andreas Klotz was born in 1968. After training as a typesetter, he decided to become self-employed at the young age of 19. Today, he is the owner and managing director of TiPP 4 GmbH, an advertising agency and publishing company in Rheinbach, near Bonn. Nature photography has been his hobby for over 30 years. Andreas is married and the father of two teenagers. Every now and then he gets „permission" to go on trekking or photography holidays for two to three weeks without his family, to get away from work and the daily routine. Then he journeys far and wide to see nature at its finest, take lots of photos and chart his experiences in travel logs. He discovered his passion for travelling in 1988 on his first photo safari in Kenya. He has been to Africa more than twenty times, and to South America, and visited India for the first time for this project.

Ingo Knoll wurde 1961 geboren. Der gelernte Bank- und Diplom-Kaufmann arbeitete nach dem Studium zunächst in der Familien-Großhandlung und machte sich später mit einem Unternehmen für Mobil- und Telekommunikation, der NAVITEL GmbH in Bayreuth, selbstständig. Parallel zu seiner Liebe für das Reisen entwickelte sich auch die Leidenschaft für die Fotografie. Nach der früheren Begeisterung für Städte- und Reisemotive liegt heute der Schwerpunkt seiner Arbeiten bei der Tierfotografie. Zuletzt unternahm er Reisen zu Alaskas Grizzly- und Polarbären sowie den Großkatzen der MasaiMara, als Nächstes stehen robbenjagende Orcas in Argentien und OrangUtans auf Borneo auf seinem Wunschzettel. Auf der Indien-Reise begeisterten ihn die freundlichen und immer fotogenen Bewohner des Subkontinentes.

Ingo Knoll was born in 1961. After studying banking and business administration, he worked for the family wholesale business and later established a mobile communications and telecommunications company, NAVITEL GmbH, in Bayreuth. His love of travelling developed in parallel with his passion for photography. Having previously concentrated on capturing images of cities and his travels, he now focuses on photographing animals. He recently travelled to Alaska to take shots of grizzlies and polar bears, and to the Maasai Mara National Reserve in Kenya for the big cats. Next, he plans to head off to Argentina to see the orcas hunting for seals and to Borneo to see the orang-utans. In India, he was particularly taken with the friendly and very photogenic people of the sub-continent.

Raghunandan Kulkarni wurde 1980 im indischen Pune geboren und arbeitet momentan als stellvertretender Geschäftsführer für Indian Wildlife Experiences. Fotografieren und die Natur waren seit seiner Kindheit Raghus Leidenschaft und er begann seine Karriere im Jahr 1995 bereits im Alter von 15 Jahren. Im Kanha-Nationalpark arbeitete er drei Jahre lang als Naturforscher, wo er sein Wissen über wilde Tiger erweiterte, sie fotografierte und auf diese Art unmittelbare Erfahrungen mit der indischen Wildnis machte. Später arbeitete er für verschiedene Firmen auf der ganzen Welt bei Reisen nach Indien in Gebiete mit Wildtieren als Tourenführer. Als weltweit gefeierter Naturfotograf, der insbesondere für seine Tigerfotos geschätzt wird, wurde Raghu im Jahr 2008 der Pu. La. Deshpande-Preis (die prestigeträchtigste Auszeichnung in Maharashtra) verliehen.

Raghunandan Kulkarni was born in 1980 in Pune in India. Currently working as an associate vice president of Indian Wildlife Experiences. Raghu has been passionate about wildlife and photography ever since he was a child and began his career at a tender age of 15 in 1995. He worked as a naturalist in Kanha national park for 3 years where he learnt more about wild tigers and photographed them and got the first-hand experience of Indian wilderness. Later on he worked as a wildlife tour leader for many companies from the world for India. A worldwide acclaimed wildlife photographer especially known for his tiger photography. Raghu has been awarded by PU. LA. Deshpande (most prestigious award in Maharashtra state) award in 2008.

Fotografen-Viten | Photographer profiles

Harry P. Lux erblickte 1965 in Trier das Licht der Welt. Seit 36 Jahren ist er immer mit Kameras & Objektiven unterwegs. Nach der Foto-Ausbildung schloss er in Berlin das Studium zum Phototechniker ab. Auf vielen Reisen durch Amerika und die meisten Länder Europas sucht er immer Tier-Begegnungen. Alles, was sich bewegt, zählt zu seinen Lieblingsmotiven. Er schreibt Praxistests für viele Zeitschriften und leitet zahlreiche Fotoworkshops in Europa und den USA. Indien und die Tiger waren seine große Sehnsucht – die Tiere & Menschen auf dieser Indienreise haben seine Erwartungen bei Weitem übertroffen. Sein Fazit: „Indien, sofort und immer - wieder". 1994 erweckte er seine eigene Publikation HarrysInFocus zum Leben und füllt heute mit Fotos, Praxistests & Erfahrungen seine lebendige Homepage: www.Harrys-Fototagebuch.de

Harry P. Lux was born in Trier in 1965. He has been travelling with cameras and lenses in his luggage for 36 years. After training as a photographer, he completed his studies in Berlin, graduating as a photo technician. He has been on many trips around America and most of Europe, always on the look-out for animals. He loves to photograph anything that moves. Harry writes product reviews for various magazines and runs numerous photography workshops in Europe and the USA. He had always wanted to see India and the tigers – and the animals and people he experienced on this trip far exceeded his expectations. Harry says he would „go back to India any time at the drop of a hat". He launched his own publication, HarrysInFocus, in 1994 and now has his own Internet site full of photos, product reviews and experiences: www.Harrys-Fototagebuch.de

Harald Lydorf, Jahrgang 1945, ist aufgewachsen in der Eifel und Trier. Nach einem Studium der Nachrichtentechnik in Aachen und einigen Zwischenstationen wohnt er mit seiner Frau in der Nähe von Bonn. Er ist auf dem Weg in den Ruhestand, aber Reisen in Naturparadiese wie Island, Colorado-Plateau in den USA, die Sahara, Namib, Kalahari sowie viele Nationalparks in Afrika und die damit verbundene Fotografie bringen genügend Abwechslung und Nervenkitzel. Selbst auf diesen „touristischen" Reisen hat er schon Veränderungen in den Beständen verschiedener Tierarten feststellen können/müssen. Deshalb ist er von der Idee des „Mondberge-Projektes" begeistert und macht zum dritten Mal mit. Seit diesem Indienbesuch sind die eleganten kraftvollen Tiger seine Favoriten bei den Großkatzen – sie sollen uns erhalten bleiben.

Harald Lydorf, born in 1945, grew up in the Eifel region and in Trier. He studied communications engineering in Aachen and, following a long career, now lives near Bonn with his wife. He is close to retiring, but trips to natural paradises such as Iceland, the Colorado Plateau in the USA, the Sahara, the Namib, the Kalahari and various national parks in Africa, together with the photography along the way, bring more than enough variety and excitement. Even on these „touristy" trips, he has seen for himself how the wild animal populations are changing. That's why he is so enthusiastic about the idea of the Mondberge-project and was on board once again for this, the third instalment. Since visiting India, the elegant and powerful tiger is now his favourite among the big cats. Their survival is imperative.

Michael Matschuck, 1964 in Krefeld geboren, lebt heute mit seiner Frau und seinen beiden Kindern in Kamp-Lintfort am Niederrhein. Er ist als geschäftsführender Gesellschafter bei der Firma druckpartner Druck- und Medienhaus in Essen tätig. Der Naturliebhaber und semi-professionelle Fotograf verbindet seine Reisen am liebsten mit dem Fotografieren und dem Schutz von bedrohten Tierarten. Er ist fester Partner des „Mondbergeprojekts" und im Vorstand der Artenschutzorganisation AGA in Stuttgart. Kommunikation und Ökologie, eine ausgeglichene und bewusste Lebensweise und der Einklang des Menschen mit der Natur sind ihm wichtig. Mit seinem Engagement für den Natur- und Artenschutz und der Arbeit bei Mondberge und Fotografeneditionen möchte er andere Menschen auch dazu bewegen.

Michael Matschuck, born in Krefeld in 1964, currently lives in Kamp-Lintfort in the Lower Rhine region with his wife and their two children. He is the managing partner of druckpartner Druck- und Medienhaus, a printing and media company in Essen. A nature lover and semi-professional photographer, he combines his travels ideally with photography and the protection of endangered species. He is permanent partner of the Mondberge-project and a member of the board of the AGA wildlife conservation organisation in Stuttgart. Subjects close to his heart include communication and ecology, a balanced and aware way of life, and harmony between mankind and nature. He hopes his commitment to the conservation of nature and wildlife and his work for Mondberge and Fotografeneditionen will encourage others to think similarly.

Martin Ruhnke, geboren 1967 in Köln, kam schon sehr früh zur Fotografie. Auf vielen Reisen mit seiner Familie machte er erste Erfahrungen mit einer AGFA-Kamera vom Vater. Schon damals lag der Focus dabei auf Tier und Natur. Mit der ersten Maledivenreise 1991 wurde die Naturfotografie schnell zu Martins Hauptbeschäftigung in der Freizeit. Bald fand sich die erste SLR-Kamera, eine analoge NIKON ein. 2005 stellte Martin auf die digitale Fotografie um. Seit den ersten Schritten mit einer SLR-Kamera ist Martin bei NIKON geblieben. Aktuell nutzt er eine NIKON D7000 und eine NIKON D80 im Hugyfot-Gehäuse für Unterwasser-Aufnahmen. Martin veröffentlicht regelmäßig Reiseartikel in verschiedenen Magazinen. Er ist auch als Reiseveranstalter tätig und bietet Reisen für Naturliebhaber und Fotografen weltweit an.

Martin Ruhnke, born in Cologne in 1967, took up photography at a very early age. His first steps were with his father's AGFA camera on various trips with his family. Even then, his attention was caught by animals and nature. Nature photography quickly became Martin's greatest hobby following his first trip to the Maldives in 1991. He soon had his first SLR camera, an analogue Nikon. Martin switched to digital photography in 2005. He has remained faithful to Nikon ever since his first experience with an SLR. He currently uses a Nikon D7000 and a Nikon D80 in a Hugyfot case for underwater shots. Martin regularly publishes articles about his travels in various magazines. He also works as a travel agent, organising trips all over the world for nature lovers and photographers.

Susanne und Peter Scheufler

Peter Scheufler wurde 1958 geboren. Nach seinem Physikstudium hat er sich 1984 selbstständig gemacht und sich mit elektronischen Steuerungen befasst. 1990 gründete er die LeuTek GmbH, die mittlerweile über 60 Mitarbeiter hat. Der begeisterte Sporttaucher und Naturliebhaber beschäftigte sich bereits während des Studiums mit der Fotografie, die immer mehr in den Vordergrund trat, nachdem er und seine langjährige Lebensgefährtin Susanne (Jahrgang 1961) 2005 bei einer Motorradtour durch Namibia ihre Liebe zu Afrika entdeckten. Die beiden kennen sich seit 1994 und waren sieben Jahre lang in einer Kunden-Lieferanten-Beziehung „per Sie", bevor sie 2001 ein  Paar wurden und Susanne von Hannover nach Stuttgart umsiedelte und auch in der LeuTek aktiv wurde. Seit 2009 sind die beiden nicht mehr operativ in der Firma tätig und können sich seitdem intensiv dem Reisen und der Fotografie widmen. Seit 2012 leben sie halbjährig in Südafrika am Kap, wo sie am 12.12.2012 heirateten, und im deutschen Sommer am Bodensee. Neben vielen Fotosafaris durch zehn Länder im südlichen Afrika, bereisten sie unter anderem Patagonien, die Arktis und die Antarktis. Sie waren nun das erste Mal in Indien und beeindruckt von den Menschen, den schillernden Farben und der Natur. Insbesondere die majestätischen Tiger haben es Ihnen angetan.

Susanne und Peter Scheufler

Peter Scheufler was born in 1958. After studying physics, in 1984 he became self-employed in the field of electronic control systems. In 1990, he founded LeuTek GmbH, which now has over 60 employees. An enthusiastic recreational diver and nature lover, he got into photography during his studies. It became more and more important to him after he and his long-term partner, Susanne (born in 1961), went on a motorcycle tour of Namibia in 2005 where they discovered their love of Africa. They have known each other since 1994 and, for the first seven years, were not even on first-name terms as they were in a customer/supplier relationship. They finally got together as a couple in 2001, after which Susanne moved from Hanover to Stuttgart and joined LeuTek. They have not been involved in the company's day-to-day operations since 2009, which means they are able to dedicate their time to travelling and photography. They now spend half the year in South Africa at the Cape, where they married on 12/12/2012, and the German summer at Lake Constance. In addition to many photo safaris around ten countries in southern Africa, they have also travelled to Patagonia, the Arctic and Antarctica. This was their first trip to India and they were impressed by the people, the bright colours and the flora and fauna. They were particularly taken by the majestic tigers.

Fotografen-Viten | Photographer profiles

Kerstin von Splényi, geboren 1965 in Leipzig, lebt bei Heidelberg. Neben ihrem Beruf als Bauingenieurin betreibt sie nach einem Zweitstudium seit 2012 eine PR-Agentur. Die Sehnsucht nach fernen Ländern ist ihr als echtem „Ostkind" in die Wiege gelegt. Verständlich, dass sie sich seither immer wieder auf das Abenteuer Fernreise einlässt. Zweites, intensiv gelebtes Hobby ist die Fotografie. Mit ihrem seit zwei Jahren erscheinenden Walldorf-Kalender, der die Schönheit ihrer Wahlheimat spiegelt, verbindet sie Fotografie mit künstlerischer Verfremdung. Gemeinsam mit ihrem Mann hat sie Europa, Afrika und Amerika bereist. Ihre unbändige Leidenschaft für Großkatzen, die 2005 in Ostafrika begann, war Antrieb für ihre zweite Indienreise. Als freie Journalistin zeichnet sie für einen Teil des Buchtextes verantwortlich.

Kerstin von Splényi, born in Leipzig in 1965, lives near Heidelberg. In addition to being a trained construction engineer, she studied for a second time before launching a public relations agency in 2012. A child of the former German Democratic Republic, she grew up with a yearning to see other countries. No wonder, then, that she regularly embarks on long-distance adventures. Her second great hobby is photography. She has been publishing her calendar of Walldorf for two years now, which combines photography with artistic alienation and reflects the beauty of her adoptive home. She has travelled around Europe, Africa and America together with her husband. Her unbridled love for big cats began in eastern Africa in 2005 and was the driving force behind her second trip to India. As a freelance journalist, she wrote some of the texts in this book.

Lothar Spranger wurde 1967 geboren. Nach seiner Ausbildung zum Raumausstatter hat er sich 1995, nach absolvierter Meisterprüfung, selbstständig gemacht. Als Inhaber einer Autosattlerei bestreitet er nunmehr so seinen Lebensunterhalt. Seit 25 Jahren ist er mit seiner Partnerin Renate zusammen, die seit 2004 auch seine Frau ist. Vor sechs Jahren hat er das Fotografieren wieder entdeckt, und ist mit der Kamera vor allem in heimischen Gefilden unterwegs. Als Hobbyfotograf war es für ihn eine interessante Erfahrung, an einer solchen Projekt-Fotoreise teilnehmen zu dürfen.

Lothar Spranger was born in 1967. After training as an interior designer and qualifying as a master craftsman, he became self-employed in 1995. He now earns his living as the owner of an automobile upholstery workshop. He has been with his partner Renate for 25 years and they have been married since 2004. He rediscovered photography six years ago and is mostly out and about with his camera in his local area. As an amateur photographer, it was an interesting experience for him to take part in a project-related photo holiday such as this.

Bildnachweise | Photo credits

Katrin und Jens Beck
20-21, 34/4, 39 m., 39 u., 46 u., 50, 55 l., 59/2, 61 u., 64/1, 65 o., 67/3, 68/3, 73, 75, 76 l., 79 o., 80/2, 80/3, 153 u., 163, 165/1, Schuber Cover o. r.

Frank Hanel
4-5, 11/4, 13, 33 o., 34/1, 34/3, 35/4, 37/2, 37/3, 37/6, 38-39, 39 o., 40 u. l., 43 o., 47/2, 48-49, 53 o., 55 r., 56 o., 72, 74, 79 u., 81 u., 83 o., 87 m., 165/3, 165/5, Schuber Rückseite o. m., Schuber Rückseite u. l.

Andreas Klotz
10/5, 16-17, 19, 92, 98 o., 98-99, 100 u., 102, 102-103, 106 u., 112, 114, 119, 120 o., 122 u., 123, 126-127, 129, 136-137 o., 140-141 o., 140 u., 141 u., 142 m., 144-145, 150 o., 157, 159 u. r.

Ingo Knoll
28-29, 35/1, 36/1, 36/2, 36/3, 36/5, 38 o., 44/2, 47/3, 53 u., 54, 56 u., 58 o., 59/3, 64/3, 69 u., 76 r., 77 r., 78 u., 80/1, 82 u., 83 u., 86 m., 86-87, 87 u., 164 o., 165/4, Schuber Rückseite u. m.

Raghunandan Kulkarni
61 o., 66 o., 66 m., 66 u., 68/1, 80/4, 86 u., 95, 115, 122 o., 124 o., 124 m., 125 r. u., 149, 151, 164 u.

Harry P. Lux
10/4, 45/4, 93, 107/1, 107/3, 133 u., 135, 136 m., 136 u., 136-137 u., 138, 139 o., 139 u., 150 u., 158 o. l., 158 u., 159 u. l., Schuber Cover u., Schuber Rückseite o. r.

Harald Lydorf
18, 24, 27, 35/5, 36/4, 37/1, 37/4, 37/5, 40 o., 40 m., 41, 44/1, 45/1, 45/2, 51, 52, 57, 58 u., 64/4, 70-71, 87 o., 152 o., 165/2, 176

Fotografen-Viten | Photographer profiles

Heike Zachenhuber, geboren1974 in Homberg (Efze), ist Betriebswirtin und arbeitet heute als Maklerbetreuerin für eine große deutsche Versicherungsgesellschaft. Als Zehnjährige bekam Sie ihren ersten Fotoapparat geschenkt. Sie lebte ein Jahr in New York, zur Dokumentation dieser Zeit kaufte sie sich ihre erste analoge Spiegelreflexkamera, machte überwiegend Schwarzweißfotos, die sie auch meistens selbst entwickelte. 2004 lernte sie ihren Mann Norbert kennen und lieben. Das gemeinsame Hobby entwickelte sich rasant weiter. Auf vielen gemeinsamen Reisen rund um die Welt legte sie den Fokus zuerst auf Land und Leute. Später rückten erst Vögel und dann immer mehr größere Tiere vor die Linse. Nach Afrika und Sri Lanka hat sich 2014 ihr Traum erfüllt: die Fotoreise für dieses Buch mit Gleichgesinnten nach Indien.

Heike Zachenhuber, born in Homberg (Efze) in 1974, studied business administration and currently works as a broker manager for a large German insurance company. She was given her first camera at the age of ten. She lived in New York for a year and bought her first analogue single-lens reflex camera to document her time there, mostly taking black and white photos which she then developed herself. She met and fell in love with her husband Norbert in 2004. Their shared hobby took off very quickly and they initially focussed on the countries and the people on their many trips to different parts of the world together. Later, they turned their attention to birds, and then, increasingly, to larger animals. Having been to Africa and Sri Lanka, another dream came true in 2014 when they joined the photo holiday for this book together with like-minded people.

Norbert Zachenhuber, geboren 1962 in München, gelernter Werkzeugmachermeister, ist heute durch den Wechsel in die Medizinproduktebranche als Vertriebsleiter tätig. Durch ein berufsbegleitendes Studium hat er nur wenig Zeit, sich seiner größten Leidenschaft, der Fotografie, zu widmen. Er fotografiert seit dem 8. Lebensjahr. Seitdem galt es für ihn, besondere Momente für die Ewigkeit auf möglichst unterschiedlichen Gebieten festzuhalten. So entstanden interessante Bilder von Menschen, Landschaften, Tiere, Vögel und Kunst im öffentlichen Raum. Immer mehr entwickelte er die Faszination für die Naturfotografie. Erst 2010 wurde endlich der langersehnte Wunsch einer Safari in Namibia, Botswana und später Sri Lanka erfüllt. Die Indien-Projektreise ist sein persönliches Highlight, hier konnte er die Fotografie mit Artenschutz verbinden.

Norbert Zachenhuber, born in Munich in 1962, trained as a master tool maker but is now a sales manager in the pharmaceutical industry. He has been taking photographs since the age of 8, and it is his greatest passion, but studying in parallel with his job means he cannot devote nearly as much time to it as he would like to. He has always been interested in capturing special moments for eternity in as broad a range of areas as possible. He has taken pictures of people, landscapes, animals, birds and art in public spaces. His fascination with nature photography grew and grew, and in 2010 he finally fulfilled his lifelong wish to go on a safari, first to Namibia, then Botswana and later to Sri Lanka. The project trip to India is his personal highlight, allowing him to combine photography with the conservation of endangered species.

Bildnachweise | Photo credits

Michael Matschuck
10/3, 59/1, 60 u., 62-63, 64/2, 65 u., 67/1, 68/2, 69 o., 77 l., 78 o., 81 o., 82 o., 84-85, 86 o., 153 o. l., 160-161, 162

Martin Ruhnke
10/2, 34/5, 60 o., 67/2, 116 o., 169

Susanne und Peter Scheufler
8-9, 10/1, 11/3, 15, 22-23, 25, 30, 32, 34/2, 35/3, 36/6, 40 u. r., 44/3, 44/4, 45/3, 46 o., 47/1, 91, 94, 97 o., 100 o., 101, 104-105, 106 o., 107/2, 108-109, 110, 111, 118, 120 u., 124 u., 125 l. u., 128, 130-131, 132, 137 o., 141 o., 142 u., 143, 144 o., 144 u., 152 u., 153 o. r., 159 o., Schuber Cover o. l., Schuber Rückseite o. l., Schuber Rückseite u. r.

Kerstin von Splényi
43 u., 90, 117, 121 o., 125 o., 133 o., 134 o., 140-141 m., 142 o. l., 146-147

Lothar Spranger
11/1, 26, 31, 88-89, 97 u., 103 o., 116 u., 134 u., 145 u., 159 m. r.

Heike und Norbert Zachenhuber
11/2, 33 u., 35/2, 42, 47/4, 96, 98 u., 113, 121 u., 139 m., 142 o. r., 148, 154-155, 156, 158 o. r.